ESTIMATING BUILDING CONSTRUCTION

QUANTITY SURVEYING

Second Edition

ESTIMATING BUILDING CONSTRUCTION

QUANTITY SURVEYING

WILLIAM J. HORNUNG

California State Polytechnic University
Pomona, California

PRENTICE-HALL, Englewood Cliffs, N.J. 07632

Library of Congress Cataloging-in-Publication Data

Hornung, William J.
  Estimating building construction.

  Includes index.
  1. Building—Estimates. I. Title.
TH435.H88 1986   692′.5   85-19203
ISBN 0-13-289919-1

Editorial/production supervision and
  interior design: *Ellen Denning*
Manufacturing buyer: *John Hall*

10 9 8 7 6 5 4 3 2

ISBN 0-13-289919-1 025

PRENTICE-HALL INTERNATIONAL (UK) LIMITED, *London*
PRENTICE-HALL OF AUSTRALIA PTY. LIMITED, *Sydney*
PRENTICE-HALL CANADA INC., *Toronto*
PRENTICE-HALL HISPANOAMERICANA, S.A., *Mexico*
PRENTICE-HALL OF INDIA PRIVATE LIMITED, *New Delhi*
PRENTICE-HALL OF JAPAN, INC., *Tokyo*
PRENTICE-HALL OF SOUTHEAST ASIA PTE. LTD., *Singapore*
EDITORA PRENTICE-HALL DO BRASIL, LTDA., *Rio de Janeiro*
WHITEHALL BOOKS LIMITED, *Wellington, New Zealand*

CONTENTS

PREFACE TO SECOND EDITION

In this text for the beginning construction estimator, the emphasis is placed on finding the quantities of building materials. However, material and installation costs have also been included in this second edition, together with a detailed description of construction methods and procedures. The costs that are given are average costs over a number of larger cities in the country. They do not necessarily reflect current or local costs and are given only for the purpose of problem solving.

Six additional units of instruction were added to the original sixteen, requiring approximately two semesters of two-hours sessions. Each study unit consists of:

Introduction. This section alerts the student to the contents of the unit and further clarifies subject content.

The Take-off. This is the main body of the text, providing detailed explanations with numerous illustrations.

Material and Installation Costs. Costs of materials discussed in the take-off section are given here, together with installation costs and an additional cost for the subcontractor's overhead and profit.

Self Examination. Here the opportunity is provided for self-testing by the student. Short classroom problems are given for which answers are supplied.

Assignment. An outside-of-classroom assignment is given after each unit. The assignment is submitted to the instructor for evaluation and correction and returned to the student during the following session.

Supplementary Information. This part is intended for those who wish to enrich their understanding of the subject in the unit.

Review Questions.   Four or five pertinent review questions are provided after each unit for classroom discussion or out-of-class assignment.

Although the text does not propose to provide answers to all estimating problems, it does provide a practical sequence for first use as a textbook for a college or university course, or for use in a technical or in-house program for industry personnel.

WILLIAM J. HORNUNG
Mission Viejo

ABBREVIATIONS

A	ampere
bbl	barrel
B.F.	board foot
B.M.	board measure
brg.	bearing
Btu/hr	British thermal units per hour
C	hundred
cem.	cement
cfm	cubic feet per minute
c.i.	cast iron
conc.	concrete
C.S.F.	hundred square feet
cu ft	cubic foot
cu yd	cubic yard
d	penny (nail size)
dia.	diameter
DWV	drain, waste, vent
ea.	each
excav.	excavation
fbm	feet board measure of lumber
fin.	finish
flr.	floor
fpm	feet per minute
ft	foot
ga.	gauge
gal	gallon
hp	horsepower
hdr.	header
HVAC	heating, ventilating, air-conditioning
in.	inch
insul.	insulation
kip	1000 pounds
lb	pound
lin ft	linear foot
L.L.	live load
M	1000
mat.	material
min.	minimum
o.c.	on center
pc.	piece
psi	pounds per square inch
sq ft	square foot
sq in	square inch
sq yd	square yard
V	volt
W	watt
Ω	ohm

INTRODUCTION

The prime purpose of this book is to show the builder and student a method of taking-off building material quantities from plans of buildings of residential, commercial, or any other type of structure. Too often, costly errors in an estimate are traced to a poorly prepared *take-off*, the itemizing of the quantities of various building materials.

To prepare an accurate take-off requires of the estimator a complete understanding of all phases of construction work. In particular, it requires an understanding of the plans, elevations, sections, and details of building construction.

The estimator in a well-organized contractor's office is in most cases the center of activity. He or she controls the job from the time the first contact is made concerning a proposed building until the job is well under way. The estimator's duties include quantity surveying, interviewing subcontractors, getting quotations on materials, and preparing the estimate.

There are usually two kinds of estimates. The preliminary estimate is generally made before the final plans and specifications are complete. Such an estimate enables the architect to find out the extent to which he or she may go in developing and in detailing final plans and specifications.

Regular estimates, on the other hand, are made by the general contractor's estimator from the completed working plans and specifications.

When construction gets under way, the estimator turns over complete plans, specifications, copies of all subcontracts, and all other available data to the superintendent, who from then on is in complete charge of the job.

THE ARCHITECT

The architect prepares the "working plans" and the specifications and represents the owner's interest at all times. In the beginning the architect deals only with the owner. After drawings and specifications have been prepared, the architect (or sometimes the owner) will invite several contractors to bid on the job. Public work must be advertised, the bids must be submitted up to a certain date, and the lowest qualified bidder must be given the job. The architect will make available one set of plans and specifications to each of the bidding contractors, and the contractors will prepare their estimates based on these data.

THE CONTRACTOR

The contractor, sometimes called "general" or "builder," is responsible for the entire job. The contractor has to obtain building permits and bonds, establish the necessary safeguards, and provide temporary facilities for management, material storage and sanitation, and a water supply.

SUBCONTRACTORS

Subcontractors are persons or firms who work for general contractors and handle the parts of the building that the general contractor either does not want to handle or cannot handle. It is the general contractor's duty to coordinate the work of all subcontractors and see to it that every "sub" is on the job at the proper time and that the various trades do not interfere with each other. The general contractor sometimes has portions of the work, such as the carpentry or the brick work, done by his or her own workers; other times the general contractor is just a coordinator of all trades.

MATERIAL PEOPLE

"Material people" is the title given to firms selling construction materials. For instance, a lumber dealer is a material person. Such a person sends the lumber to the job but does not supply labor (carpenters) to erect the work.

MECHANICS

Mechanics are people who work with tools. Some trades employ apprentices and helpers, who also may be classified as laborers.

GENERAL CONDITIONS

The general conditions of the contract consist of 14 articles that define rules, regulations, and general practices pertaining to all those concerned with building construction work. These rules are normally applicable to all construction projects, with portions of them expanded or voided to fit the conditions of the specifications. The subjects covered by the articles are generally as follows:

1. Contract documents
2. Architect
3. Owner
4. Contractor
5. Subcontractor
6. Work by owner or by separate contractor
7. Miscellaneous provisions
8. Time
9. Payments and completion
10. Protection of persons and property
11. Insurance
12. Changes in work
13. Uncovering and correction of work
14. Termination of the contract

SPECIFICATIONS

Specifications guide the whole construction job as to quality of materials, workmanship, and relations between the parties concerned with the job. A good specification should be written in the same sequence as the various trades start working on the job, starting with the general conditions and continuing with demolition, clearing site, excavation, and so on.

The specifications also provide that the general contractor and subcontractors comply with all local, state, and government rules, regulations, and ordinances; file all necessary documents and information; and pay for and obtain all licenses, permits, and certificates of inspection.

Following is a sample specification written for an excavation contractor. This specification is not for any of the plans in the text.

EXCAVATION, BACKFILL, AND GRADING

General Conditions

The general conditions of the contract of the American Institute of Architects, current edition, shall form a part of this division, together with the special conditions, to which this contractor is referred.

Scope

The work under this division includes all labor, material, equipment, and appliances required for the complete execution of all excavation, backfill, and grading, as shown on the scale drawings, or as may be reasonably inferred to make this division complete, which is generally as follows:

1. General excavation
2. Hand excavation
3. Drain trench excavation
4. Backfill around walls and under floors
5. Excavation and backfill for septic tank
6. Rough grading

It is to be assumed that the contractor shall include all labor, materials, equipment, and appliances necessary to complete the work, whether specifically mentioned in the specifications or not, except only those items specifically mentioned as omitted from this division of the specification.

Work Not Included

The following items will be performed under other divisions of this contract, separate contracts, and future work, and will not be included as part of this division:

1. Finish grading
2. Landscaping

General Excavation

The contractor shall excavate under concrete slab for stone fill, for foundation walls and footings, and under stone fill of asphalt-paved areas of the building, to the dimensions indicated on the plans.

The contractor shall make all required allowances beyond the foundation walls for the sloping banks, sheet piling, if necessary, and for forms for foundations and waterproofing of foundation walls.

Rock excavation is not contemplated in the contract work. Quote unit price for removal of rock encountered in excavation over 27 cubic feet. This unit price is to include the removal of the material from the site.

Hand Excavation

Excavation for the footings and for the chimney footing. Excavation shall be made wide enough to allow for the installation of forms for footings.

Drain Trench Excavation

Where terra-cotta drains to septic tank are shown on the plans, the contractor shall do all excavation of trenches required for their installation. All trenches shall be 3 ft below the lowest finished grade level and shall slope toward the septic tank.

Excavation For Septic Tank and Leaching Pool

The contractor shall excavate for the septic tank and leaching pool. The excavation of the septic tank shall be 4 ft below the house sewer inlet. The width of the excavation shall be 6'6" and the length 9 ft.

The excavation for the leaching pool shall be 5 ft below the inlet and shall be dug to a diameter of 5'8".

After the septic tank and leaching pool have been constructed by others, the contractor shall backfill all trenches and areas around septic tank and leaching pool up to an elevation of the present grade.

Backfill Around Walls and Under Floors

After the footings, foundations, and retaining walls are built and the other work required on the exterior of the walls is completed, the contractor shall backfill carefully around these walls with good clean earth up to the level and grades indicated on the plans.

The backfill under floors shall be spread, sprinkled, and tamped or rolled to a solid and compact condition, ready to receive the cement floor construction, installed by others.

Rough Grading

All excavated earth not used for backfill shall be spread around the building to the level or levels to within 6 in. of the finished grade and 10 ft from the building line.

A TYPICAL TAKE-OFF

Figure I-1 is a typical take-off of excavations for the removal of top soil, general excavations for basements or cellars, and the final 6-in. footing excavation to undisturbed soil.

The take-off is prepared on typical column-type ruled paper, listing the items,

EXCAVATION					
Item	Units	Length (ft)	Width (ft)	Depth (ft)	Cubic Feet	Cubic Yards
Remove Topsoil						
	1	174.50	81.50	.50	7111.00	263.30
General Excavation						
Main building	1	50.83	33.00	7.50	12580.00	
Central storage	1	23.67	26.75	8.50	5382.00	
Boiler room	1	10.67	25.75	9.50	2610.00	
Outside stair	1	18.16	5.05	8.50	810.00	
					21382.00	792.00
Footing excavation						
Footing under walls	2	46.00	2.33	.50	127.00	
	2	23.33	2.33	.50	54.35	
Extra for chimney	1	6.33	1.00	.50	3.17	
Hall stair	2	23.33	2.00	.50	46.66	
Kitchen and pantry	1	14.00	2.00	.50	14.00	
Central storage	2	21.75	2.00	.50	43.50	
	2	14.67	2.00	.50	29.34	
Boiler room	2	15.67	2.00	.50	31.34	
Men's locker room	1	37.67	2.00	2.50	188.35	
	1	19.00	2.00	2.50	95.00	
	1	28.50	2.00	2.50	142.50	
	1	8.75	2.00	2.50	43.75	
Outside stair	1	16.67	2.00	.50	17.67	
	1	1.25	2.00	.50	1.25	
Ladies locker room	2	33.33	2.00	2.50	333.30	
	2	16.67	2.00	2.50	166.70	
	2	15.67	2.00	2.50	156.70	
Front porch	1	44.00	2.00	2.50	220.00	
	2	8.00	2.00	2.50	80.00	
Rear areas	1	33.50	1.00	2.00	67.00	
East area	1	8.25	1.00	2.00	16.50	
					1878.08	69.55

Figure I–1

units, length, width, and depth in feet, multiplied to find the cubic feet, and divided by 27 to arrive at cubic yards.

The dimensions are read directly from the plans, and the estimator is required to transpose the inches on the plans to decimal parts of a foot on the takeoff sheet. The length dimension of topsoil removal on the plans is 174′6″. On the take-off sheet, the dimension is written as a decimal number, such as 174.5 ft.

It becomes a relatively simple matter for the estimator to memorize the decimal equivalents of inches, for 6 in. is
6/12 of a foot, or 6.0/12 = 0.5 ft.

Inches	1	2	3	4	5	6	7	8	9	10	11
Feet	0.08	0.17	0.25	0.33	0.42	0.50	0.58	0.67	0.75	0.83	0.91

It is essential when listing the figures on the take-off sheet to write the figures within the vertical guidelines. Multiplications and divisions can quickly be achieved with the aid of a calculator.

1

EXCAVATION, BACKFILL, AND GRADING

INTRODUCTION

This unit is concerned with excavations, such as the removal of topsoil, general excavation for cellars or basements, and excavations for trenches and septic tanks.

In removing topsoil the contractor is required to excavate the rich top layer of soil to a depth of 6 in. and is usually stated in the specification, and store it on the site so that it can be used later for the final topping. The soil is figured in cubic yards. Specifications generally give the dimensions beyond extreme exterior walls of the proposed building to which the topsoil is to be removed.

The bulldozer and front-end loader (Fig. 1-1) are often used for shallow excavations such as for topsoil removal and general excavations for cellars or basements and can load directly into trucks to be hauled away. Bulldozers and front-end loaders may be designed with a backhoe at their other end. The power shovel is used for very large excavations because it is an economical method of excavating and quick loading of trucks.

Backfill and Grading

Backfill is replacing the material around foundations, footings, under slabs, against basement walls, and the filling up of trenches. The equipment used for these operations may include power shovels, backhoes, bulldozers, or other equipment.

Figure 1–1   (Courtesy of Ford Motor Company.)

Compaction of backfill is done by hand-operated air-powered compactors or with the use of rollers.

Grading is the final leveling of the grade according to the dimensions given on the plans.   This may be extensive work or may be very limited, depending on the contours of the site.

Cut and Fill

All building plans indicate the level of the finished grade in relation to the first-floor level, known as elevation 0'-0".   All height dimensions above elevation 0'-0" are called plus dimensions and those below are called minus dimensions.

In grading the land for the building, a benchmark or datum line is established as a reading point above sea level.   From this point other benchmarks may be made.   The symbol used is ⊼, where the center of the horizontal bar is the datum or *benchmark*.   A curbstone may be a benchmark in an urban area, or a wooden post protruding slightly from the ground and fenced around to keep it protected during operations may be used in a rural area.

Estimating Cut

When land is leveled for building, a *station* is any point on the land for which a recording is made in relation to the benchmark. Stations are recorded in feet and tenths of feet. They may be above, below, or directly in line with the benchmark. The benchmark is often shown as 100.0 ft, which means that all elevations taken from this figure are shown, such as 14.6 ft instead of 114.6 ft. The 14.6 ft being above 100 ft means that the earth needs to be cut 14.6 ft. An 87.6-ft dimension means a 12.4-ft fill.

Problem

Estimate the cubic yards to be cut from a parcel of land 75 ft × 75 ft (Fig. 1-2) with the station elevations as shown above the 100-ft benchmark.

Solution

Step 1: From the benchmark estimate the average height of the station.

$$\frac{14.6 \text{ ft} + 12.2 \text{ ft} + 2.8 \text{ ft} + 4.4 \text{ ft}}{4} = 8.5 \text{ ft average height}$$

Step 2: Multiply the land area by the average height of the station in feet.

$$5625 \text{ sq ft} \times 8.5 \text{ ft} = 47{,}812.5 \text{ cu ft} \div 27 = 1771 \text{ cu yd to be cut}$$

Problem

Estimate the cubic yards of earth to be cut and hauled from two adjacent parcels of land, A and B, as shown in Fig. 1-3. Assume a 100-ft benchmark.

Solution   Find the average depth of each parcel of land separately, and multiply by the areas of A and B. The average height of the station in A is 5.5 ft. Therefore,

$$5.5 \text{ ft} \times 1800 \text{ sq ft} = 9900 \text{ cu ft} \div 27 = 367 \text{ cu yd}$$

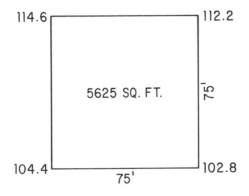

Figure 1–2

The average height of the station in B is 19.4 ft. Therefore,

$$19.4 \text{ ft} \times 500 \text{ sq ft} = 9700 \text{ cu ft} \div 27 = 359 \text{ cu yd}$$

$$\text{Total cut} = 367 + 359 = 726 \text{ cu yd}$$

Excavation for Septic Tank

A septic tank is a concrete tank embedded in the soil, into which sewage is allowed to drain. Areas without public sewers require septic tanks for each building. Septic tanks are ready made and are available in many sizes, serving from 1 to 4 persons up to 45 to 50 persons.

A tank for 1 to 4 persons is 5'0" × 2'6" with a liquid depth of 3'6" plus a 1'0" air space.

A tank for 45 to 50 persons is 11'6" × 5'9" with a liquid depth of 5'0" plus a 1'3" air space.

Excavation for tanks are made according to the size of the tank, with an allowable working space of 6 to 12 in. around the tank.

To estimate the earth excavation for the tank, multiply the length by the width by the depth of the tank, plus the working space around the tank. Divide the cubic feet by 27 to get the total cubic yards.

THE TAKE-OFF

Topsoil

For example, excavate the topsoil for the plot shown in Fig. 1-4. Assume a distance of 20'0" beyond the exterior walls of the building. (*Note:* The topsoil is removed before the proposed building is constructed.)

The length of the excavation is 100 ft; the width is 70 ft. Multiply 100 ft × 70 ft = 7000 sq ft. Multiply by 0.5 ft (6 in.), the depth of the excavation: 0.5 ft × 7000 sq ft = 3500 cu ft. Then 3500 cu ft ÷ 27 cu ft = 129.6, or 130, cu yd.

Another method of taking off this quantity is in tabular form. It is simple, and errors are less likely to occur. Figure 1-5 illustrates this method. Under the table heading "Item" the type of excavation is indicated. Under the heading "Unit" is the number of rectangular units to be excavated. In the case of an ell-shaped plan, where both legs of the ell are equal in area, two units would be indicated. The units multiplied by the length, width, and depth yield the number of cubic feet. Divide by 27 to find the cubic yards to be excavated.

General Excavation

Under this heading the contractor is required to excavate the large areas of a cellar or basement. This type of excavation is done with the aid of heavy mechanical

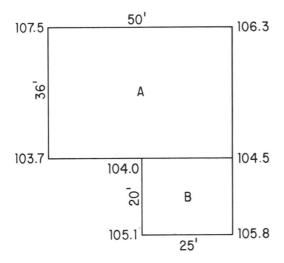

Figure 1-3

equipment such as a power shovel, bulldozer, lifting crane, or clamshell. From the plans, the estimator can determine how far below the grade level the excavation is to be carried. The contractor must make all required allowances beyond the foundation walls and retaining walls for the sloping of banks, forms for foundations, and waterproofing of the foundation walls.

To find the number of cubic yards of earth excavation for a basement or cellar, multiply the length by the width by the depth of the excavation. The result will be in cubic feet. To find the cubic yards, divide by 27.

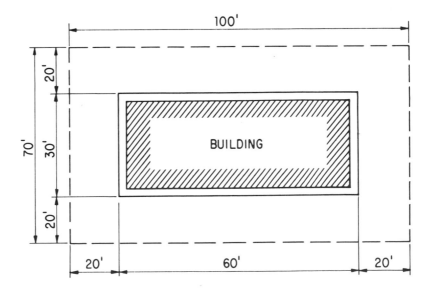

Figure 1-4

EXCAVATE TOP SOIL (20' BEYOND EXT. WALLS)						
Item	Unit	Length	Width	Depth	Cu. Ft.	Cu. Yd.
Top soil	1	100	70	0.5	3500	130

Figure 1–5

For example, given the length 50'0", the width 35'0", and the depth 7'0", find the cubic yards of excavation in Fig. 1-6. Multiply 50 ft × 35 ft × 7 ft = 12,250 cu. ft. Then 12,250 cu ft ÷ 27 cu ft = 454 cu yd. *Note:* The excavation was carried 5'0" beyond the exterior faces of walls to allow for the building of forms for the foundation walls, and for waterproofing the walls if required.

In this example, the drawing indicates that 5 ft of additional excavation is necessary. However, in most cases the estimator must choose this dimension based on knowledge of the soil type and the work to be accomplished. The appropriate slope may be determined by using Fig. 1-7.

Earth Swell and Earth Shrinkage

When earth and rock are loosened during excavation they assume a larger volume. This increase in volume is described as *swell* and is usually expressed as a percent gain compared to the original volume. If earth is placed in a fill and compacted with modern equipment, it usually occupies a smaller volume than in its natural state in the cut or borrow pit. This decrease in volume is described as *shrinkage* and is expressed as a percent of the original volume. Figure 1-8 indicates the

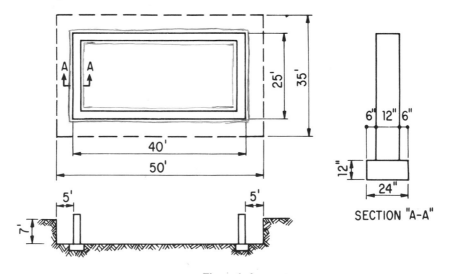

Figure 1–6

SLOPES FOR VARIOUS TYPES OF BANKED EARTH			
Earth Type	Slope of Repose		Angle of Repose
	Length	Height (ft)	
Clean sand	1'6"	1	33°41'
Sand and clay	1'4"	1	36°53'
Bottom rock	1'4"	1	36°64'
Damp soft clay	3'0"	1	26°34'

Figure 1–7

percent of swell and shrinkage for various soils. Swell and shrinkage percentages are applied when it is necessary to haul the earth by trucks either to or from the job.

An 150-cu yd excavation of cut loam would swell 20%, making 180 cu yd to be hauled away. An 8-cu yd truck would therefore make 23 trips.

If, on the other hand, 150 cu yd had to be filled, a shrinkage allowance of about 17% would be required, making a total fill-in of 175.5 cu yd.

Examples of General Excavation Take-Off for Cellar

The plan shown in Fig. 1-9 is intended as a practice take-off of earth excavation for a cellar. The dimensions given are those of the foundation walls. The detail drawing of the footing indicates how far beyond the walls the excavation is made. The irregular plan should be marked off into rectangular areas, as shown by the diagonal lines. Each area is taken-off separately.

PERCENTAGE OF EARTH SWELL AND SHRINKAGE		
Material	Swell (%)	Shrinkage (%)
Sand or gravel	14 – 16	12 – 14
Loam	20	17
Common Earth	25	20
Dense clay	33	25
Solid rock	50 – 75	–

Figure 1–8

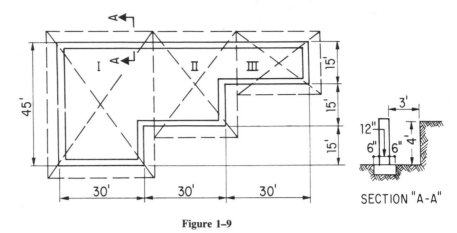

Figure 1–9

For the take-off in Fig. 1-10, carefully check each figure and verify the quantities.

Trench Excavation

In finding the cubic yards of earth excavation for trenches, multiply the length of the trench by the width by the average depth to get the cubic feet.   Divide by 27 to find the total cubic yards.

For example, Fig.1-11 shows a line of drain tile leading from a building to a dry well.   The length is 75'0".   The drain tile is pitched $\frac{1}{4}$ in./ft length.   Find the cubic yards of earth excavation.

If the drain tile is pitched, the trench for the tile is similarly pitched.   If the trench deepens $\frac{1}{4}$ in.   for every foot of length, it is obvious that in 75 ft there is $\frac{75}{4}$ or $18\frac{3}{4}$ in., which is $1'6\frac{3}{4}"$ of additional depth at the end of the trench.   If the trench at the building is 3'0" deep, the other end will be $3' + 1'6\frac{3}{4}" (18\frac{3}{4}") = 4'6\frac{3}{4}"$ deep (see Fig. 1-12).   Once the average depth is known, multiply the length

CELLAR EXCAVATION (3'-0") BEYOND EXT. WALLS						
Item	U	L	W	D	Cu. Ft.	Cu. Yd.
Area I	1	51	36	4	7344	
Area II	1	36	30	4	4320	
Area III	1	30	21	4	2520	___
					14,184	525.3

Figure 1–10

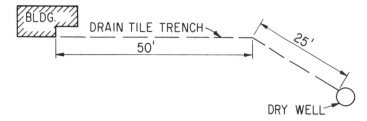

Figure 1–11

of the trench by the width by the average depth to get the cubic feet. Divide the cubic feet by 27 to get the cubic yards.

Problem

A trench is 224 ft long and is pitched $\frac{1}{8}$ in./ft length. The depth of the trench at the building is 5'0". The width of the trench is 4'6". Find the total cubic yards of earth excavation.

Solution 224-ft length at $\frac{1}{8}$-in. pitch: $\frac{224}{8} = 28"$ or 2'4". Therefore, 5' + 2'4" = 7'4" at the deep end and average depth = 5' + 1'2" = 6'2". Following is the take-off in tabular form.

TRENCH EXCAVATION, $\frac{1}{8}$ PITCH

Item	Unit	Length	Width	Depth	Cubic feet	Cubic yards
Trench	1	224	4.5	6.17	6219.36	231
				6.15	6199.2	229.6

Hand Excavation

Hand excavation, or shoveling, should be kept to a minimum, for it becomes quite expensive. However, on every job, some hand excavation is necessary.

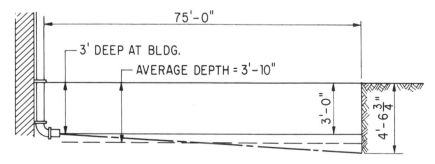

AVERAGE DEPTH OF TRENCH IS $3' + 9\frac{3}{8}" = 3'-9\frac{3}{8}"$ or 3'-10"

Figure 1–12

When the soil is of the proper consistency, the footing trenches for small buildings may be hand-excavated by spading out the exact shape of the footing. Under these conditions forms for footings are not required. To figure the earth to be excavated by hand for the footings shown in Fig. 1-6, multiply the total length of the footings by the width and by the depth. The result is so many cubic feet. To find the cubic yards, divide by 27. The cubic yards of earth excavation for footings under these conditions also equals the cubic yards of concrete required for the footings.

For example, find the cubic yards of each excavation for the footings in Fig. 1-6.

Following is the take-off of earth excavation for the footings in Fig. 1-6.

Item	Unit	Length	Width	Depth	Cubic feet	Cubic yards
Footings						
North and south	2	41.0	2.00	1.0	164	
East and west	2	22.0	2.00	1.0	88	
					252	9.3

Examples of Hand Excavation Take-Offs for Footings

The plan in Fig. 1-13 is intended as a practice take-off of earth excavation for the footing trenches. The dimensions given on the plan are those of the foundation walls. Since the footings protrude 6 in. beyond the walls, the length of the footing trench for the 60'0" wall will be 6 in. longer on both ends, making a total length of 61'0".

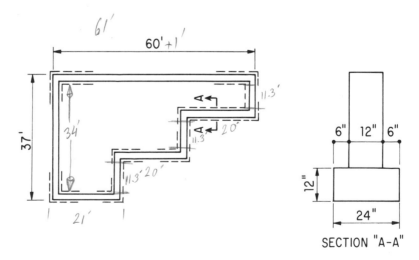

SECTION "A-A"

Figure 1–13

In this case no footing forms are required since the firm earth takes the place of the forms. Study the following take-off for the footing trenches in Fig. 1-13 and verify the figures.

FOOTING TRENCH EXCAVATION

Item	Unit	Length	Width	Depth	Cubic feet	Cubic yards
Trenches						
North and south	2	61	2	1	244	
East and west	2	34	2	1	136	
					380	14

MATERIAL AND INSTALLATION COSTS

The following equipment and labor rates are averages of some of the larger cities in the United States. They are intended only for problem solving in this book and should not be used for present-day job conditions.

Topsoil: To remove a 6-in. layer of topsoil and haul it about 200 ft using a 200-hp bulldozer will cost $0.16 per square yard. One dozer rental and operator costs $398 per day.

Excavation of continuous wall and footing: The excavation is to be 4 ft wide by 4 ft deep. A $\frac{3}{8}$ cu yd tractor backhoe can excavate 170 lin ft per day, at a cost of $2.90 per cubic yard.

General excavation: A 75-hp dozer can excavate 50 cu yd per hour and haul the material up to 50 ft at a cost of $1.02 per cubic yard.

Backfilling: A 200-hp dozer can backfill 1200 cu yd at $0.69 per cubic yd. A 200-hp dozer can backfill 900 cu of trench at a cost of $0.92 per cubic yard.

Compaction of backfill: Air tamping of light soil in 12-in. yd layers costs $11.36 per cubic yard.

Fine grading: Three passes with a motor grader can prepare 1600 sq yd per day at a cost of $0.57 per square yard.

Hand excavation: One laborer can hand shovel and load in a truck 8 cu yd per day at a cost of $11.15 per hour.

Septic tank: A 1000-gal tank size costs $300, including excavation and backfilling.

Drainage field: $\frac{1}{2}$-in open-joint 4-in.-diameter vitrified clay sewer pipe costs $3.39 per linear foot. A total of 265 lin ft of tile can be placed in one day.

The estimator must be familiar with the various pieces of equipment used in excavating, backfilling, and grading, as well as know the amount of material such equipment can excavate per 8-hr day. A good reference is *Building Construction Cost Data* by Robt. Snow Means Co., Inc., © 1984

SELF EXAMINATION

In tabular form, take-off the following quantities from the plan shown in Fig. 1-14.

1. Find the cubic yards of topsoil to be removed over the entire lot.   The depth of topsoil excavation is 6 in.
2. Using a 200-hp dozer, calculate the time required to do the job.
3. What would be the cost for this removal?
4. Find the cubic yards of general earth excavation.
5. Find the cubic yards of footing-trench hand excavation, the time required, and the cost.

Check your answers with those given in the Answer Box.

Answer Box	
Topsoil	187 cu yd
Time	2.49 hr
Cost	$124.12
Earth excavation	461 cu yd
Trenches	19 cu yd 2.375 days $211.85

ASSIGNMENT—1

1. On $8\frac{1}{2}$ in. × 11 in. white paper (similar to typing paper) sketch the plan and section shown in Fig. 1-15 on the upper half of the sheet.   Do this as carefully and neatly as you can.   On the lower half, in tabular form, take-off the quantities of earth excavation for the following: (a) general excavation for cellar and (b) footing trenches.
2. On a separate sheet of paper estimate the cubic yards of earth excavation for a drain tile trench 96 ft long, pitched $\frac{1}{4}$ in./ft length.   The depth of the trench at the building is 4′ 0″.   The width of the trench is 1′-8″.   Prepare a simple sketch, as in Fig. 1-11, and take-off the quantity in tabular form.
3. What is the time required for a tractor backhoe with a capacity of $\frac{3}{8}$ cu yd to excavate the trench for the footing?

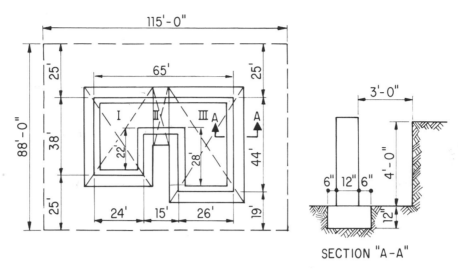

Figure 1–14

SUPPLEMENTARY INFORMATION

Volume of Trench Excavation in Cubic Yards per 50 lin ft

The following table giving various depths and trench widths will yield the cubic yards of each to be removed for every 50 lin ft.   Find the width of the trench across the top of the table and then the depth of the trench along the left side of the table.   For a trench width of 2′0″ and a depth of 1′6″, the excavated material is 11 cu yd per 100 ft of length.

CUBIC YARDS OF TRENCH EXCAVATION PER 50-FT LENGTH

Depth of trench	Width of trench						
	2′0″	2′6′	3′0″	3′6″	4′0″	4′6″	5′0″
1′0″	3.7	4.8	5.5	6.5	7.4	8.3	9.3
1′6″	5.5	6.9	8.3	9.7	11.1	12.5	13.9
2′0″	7.4	9.3	11.1	13	14.8	16.7	18.5
2′6″	9.3	11.6	13.9	16.2	18.5	20.8	23.1
3′0″	11.1	13.9	16.5	19.4	22.2	25	27.8
3′6″	13	16.2	19.4	22.7	25.9	29.1	32.4
4′0″	14.8	18.5	22.2	25.9	29.6	33.3	37
4′6″	16.7	20.8	25	29.2	33.3	37.5	41.7
5′0″	18.5	23.1	27.8	32.4	37	41.7	46.3
5′6″	20.4	25.5	30.6	35.6	40.7	45.8	50.9
6′0″	22.2	27.8	33.3	39.8	44.4	50	55.6
6′6″	24.1	30.1	36.1	42.1	48.1	54.2	60.2

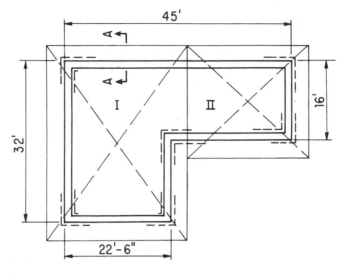

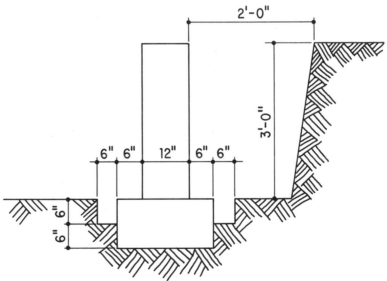

SECTION A-A

Figure 1–15

MINIMUM TRENCH WIDTHS

If depth is:	Minimum width is:	If pipe is:	Minimum width is:
1'0"	1'4"	4'–8"	1'9"
2'0"	1'5"	12"	2'6"
3'0"	1'6"	15"	2'9"
4'0"	1'8"	18"	3'0"
5'0"	1'10"	24"	3'9"
6'0"	2'0"	30"	4'3"

REVIEW QUESTIONS

1. What is the cost to remove 475 cu yd of topsoil, to what depth is the topsoil excavated, and what mechanical equipment is used?
2. Briefly explain how general excavation is accomplished and what equipment is used.
3. In excavating common soil, what is the swell and shrinkage percentage? Give the angle of repose for damp soft clay.
4. What are a "benchmark" and a "station" in land leveling?
5. Under what conditions is hand excavation used?

2

CONCRETE FOUNDATIONS

INTRODUCTION

This unit concerns itself with the quantity take-off of concrete required for footings, foundations, and other concrete generally below the first floor. The three methods of taking-off concrete—the unit method, the perimeter method, and the centerline method—are illustrated, and sample problems are provided.

Before the three methods of concrete take-offs are considered, it might be well to point out some important features of concrete. In describing the character of fresh concrete, three terms are often used: consistency, plasticity, and workability.

Consistency is a general term that describes the state of fluidity of the mix. It includes the entire range of fluidity from the driest possible mixture to the wettest.

Plasticity is used to describe a consistency of concrete that can be readily molded but which permits the fresh concrete to change form slowly when the mold is removed.

Workability describes the specific concrete mix, such as a fairly dry mix where larger aggregates can be used for a thin wall with reinforcings, yet each concrete mix has the correct workability to meet the requirements of its use in the area in which it is to be placed.

The *design of concrete mixtures* is based on the properties of the cement and the amount of mixing water used. For mixtures using sound, clean aggregates (sand and gravel) the strength and other desirable properties of concrete, under given job conditions, are governed by the proportions of mixing water to the amount of cement.

Ready-mix concrete is mostly for commercial and industrial use and is delivered to the job by the concrete company. The proper proportions of water, cement, sand, gravel, and other additions are in the ready-mix as specified.

A *field batching plant*, for very large projects, is built near the construction site. The ingredients are loaded into large metal hoppers in the proportions required for each mix, are weighed, and are delivered to the batch mixer, where measured water is added. This is to control the quality so that each batch will achieve the same strength in pounds per square inch (psi).

Concrete-mix by volume is also used widely, especially on small jobs and where a ready-mix plant is too far removed. To mix concrete directly on the job using a batch mixer of 3, 6, 11, or 16 cu yd of wet concrete, it is necessary first to determine the volume of each of the dry materials: cement, sand, and gravel.

It has been found that to place 1 cu yd of wet concrete, it will require, by volume, nearly $1\frac{1}{2}$ cu yd of dry materials. Therefore, in all estimates for volume mix, add half as much more material in the dry state by volume than the capacity of the finished volume of concrete.

For example, a concrete mix of 1:2:3 and 5 gal of water, making approximately a 3200 psi (see Fig. 2-1) concrete, is called for in the specifications. Find the total number of dry units in cubic feet.

A floor area of 10 ft × 35.5 ft × 0.33 ft (4 in.) = 118 cu ft. To the total dry mixture, add one-half extra.

$$118 \times 1.5 = 177 \text{ cu ft}$$

Then the total number of dry units required is

$$1 + 2 + 3 = 6$$

1. Cement:

$$177 \div 6 = 29.5 \text{ cu ft}$$

It requires 2 times as much sand.

2. Sand:

$$29.5 \times 2 = 59 \text{ cu ft}$$

It requires 3 times as much gravel.

3. Gravel:

$$29.5 \times 3 = 88.5 \text{ cu ft}$$

The total is 177 cu ft.

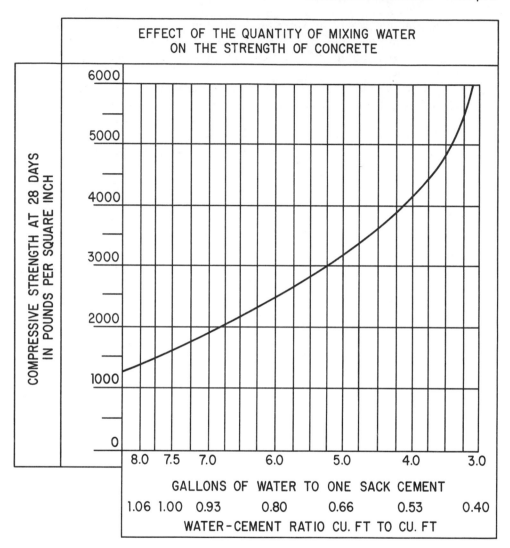

EFFECT OF THE QUANTITY OF MIXING WATER
ON THE STRENGTH OF CONCRETE

Figure 2–1

THE TAKE-OFF

To find the cubic yards of concrete wall, it is necessary to multiply the length by the width by the height of the wall in feet to find the cubic feet. The cubic feet are then divided by 27 to get the cubic yards (27 cu ft equal 1 cu yd). Assume a plan for a concrete foundation wall of a building 20 ft long, 10 ft wide, and with a 1-ft wall thickness (Fig. 2-2). How do we find the total stretch-out or the total length of wall?

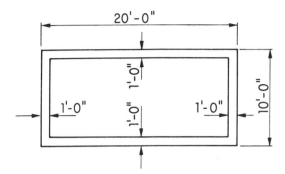

Figure 2–2

Unit Method

If we add 20 ft + 10 ft + 20 ft + 10 ft = 60 ft, this represents the outside perimeter, but it does not represent the true stretch-out of the walls.

Look at the second illustration in Fig. 2-3. Here we add 20 ft + 8 ft + 8 ft = 56 ft, the actual and correct stretch-out of the walls. In taking the two 20-ft lengths, we automatically take away from the 10-ft lengths an amount equal to the thickness of the wall. In Fig. 2-4 the shaded parts are the 20-ft lengths. To find the other two lengths, it is necessary to subtract the thickness of the wall at both ends of the 10-ft wall. Now we have two units of wall 20 ft long and two units of wall 8 ft long. Therefore,

$$2 \times 20 \text{ ft} = 40 \text{ ft}$$
$$2 \times \ \ 8 \text{ ft} = \underline{16 \text{ ft}}$$
$$56 \text{ ft of wall stretch-out}$$

Concrete Take-Off by the Unit Method

The unit method generally implies taking off the concrete in tabular form. If two walls or footings, for example, are alike in length, height, and thickness, they are

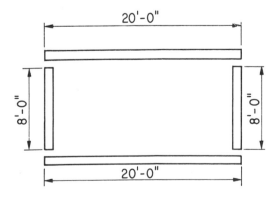

Figure 2–3

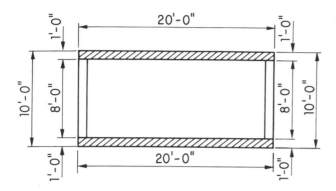

Figure 2–4

known as two units. Look at the plan in Fig. 2-5, for example, and the tabular take-off of the footings and walls by the unit method. The first two units indicated on the tabular take-off are those of the north and south footing on the plan of Fig. 2-5. Since the footing protrudes 6 in. on either end of the wall, 1 ft is added to the 20-ft length, making a total length of 21 ft. The width of the footing is 2 ft, and the depth is 1 ft. Multiply the units × length × width × depth, which equals cubic feet. The east and west footing is then computed. From the 15-ft dimension, subtract 1'6" on either end of a total of 3'0", making the length of the footing 12'0".

The foundation wall is found similarly. The first two units are those of the north and south walls shown on the plan. The length is 20'0", the width is 1'0", and the height is 6'0". Multiply these figures to get the cubic feet. The east and west wall is then computed. From the 15'0" dimension, subtract the thickness of

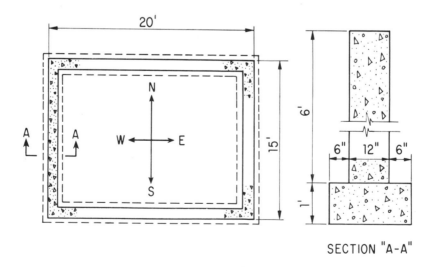

SECTION "A-A"

Figure 2–5

the wall on either end, a total of 2'0", making a 13'0" length. The width of the wall is 1'0", and the height is 6'0". Multiply these figures to get the cubic feet. Divide by 27 to get the cubic yards for the footing, since this is poured separately from the wall. Divide by 27 to get the cubic yard of concrete wall.

Concrete Take-Off by the Perimeter Method

In the perimeter method, add all the outside dimensions of the wall or footing and subtract four times the width of the wall or footing.

For example, find the cubic yards of concrete for the footing and foundation wall shown in Fig. 2-5.

The perimeter of the footing is 21 ft + 16 ft + 21 ft + 16 ft = 74 ft outside perimeter. From 74 ft subtract four times the width of the footing: 4 × 2 ft = 8 ft. Therefore, 74 ft − 8 ft = 66 ft (stretch-out of footing). The wall is found similarly: 20 ft + 15 ft + 20 ft + 15 ft = 70 ft outside perimeter. From 70 ft subtract four times the width of the wall: 4 × 1 ft = 4 ft. Therefore, 70 ft − 4 ft = 66 ft (stretch-out of wall).

To visualize the above more clearly, look at the illustration in Fig. 2-6, an enlarged corner of the plan. Note that portion A was included twice in the dimension: first in the 20-ft dimension and then in the 15-ft dimension. Four corners then require the subtraction of 4 ft from the wall and 8 ft from the footing.

The following is the take-off of the footing and foundation wall by the perimeter method for the plan shown in Fig. 2-5.

UNIT METHOD TAKE-OFF

Item	Unit	Length	Width	Depth	Cubic feet	Cubic yards
Footing	2	21	2	1	84	
North and south						
East and west	2	12	2	1	48	
					132	4.88
Wall	2	20	1	6	240	
North and south						
East and west	2	13	1	6	156	
					396	14.6

PERIMETER-METHOD TAKE-OFF

Item	Unit	Length	Width	Depth	Cubic feet	Cubic yards
Footing	1	66	2	1	132	4.88
Wall	1	66	1	6	396	14.6

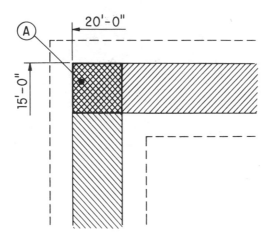

Figure 2–6

Concrete Take-Off by the Centerline Method

Another quick and easy method of finding the stretch-out of footings and walls is to assume a centerline all around the center of the walls or footings. The total of the centerlines around the building is the total stretch-out.

For example, in the plan shown in Fig. 2-5 the centerline of the 20-ft wall is 19 ft. The centerline of the 15-ft wall is 14 ft. Therefore, 19 ft + 14 ft + 19 ft + 14 ft = 66 ft of wall stretch-out. The footing stretch-out is the same: 19 ft + 14 ft + 19 ft + 14 ft = 66 ft of footing stretch-out. To understand this clearly, look at Fig. 2-7. In taking the 19 ft and the 14 ft dimension (Fig. 2-7), corner B (double shading) was included in both dimensions, whereas corner A (not shaded) was not. Therefore, the extra portion at B is substituted for portion A.

The following is the take-off of the footing and foundation wall by the centerline method for the plan shown in Fig. 2-5.

CENTERLINE-METHOD TAKE-OFF

	Unit	Length	Width	Depth	Cubic feet	Cubic yards
Footing	1	66	2	1	132	4.88
Wall	1	66	1	6	396	14.6

Worked-Out Sample Problems by the Unit, Perimeter, and Centerline Methods

Unit method (Fig. 2-8):

$$2 \times 18 \text{ ft} = 36 \text{ ft}$$
$$2 \times 12 \text{ ft} = \underline{24 \text{ ft}}$$
$$60 \text{ ft}$$

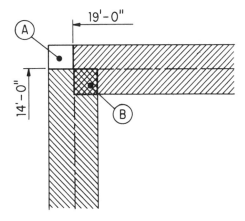

Figure 2–7

Perimeter method (Fig. 2-8):

$$18 \text{ ft} + 14 \text{ ft} + 18 \text{ ft} + 14 \text{ ft} - 64 \text{ ft} - 4 \text{ ft} = 60 \text{ ft}$$

Centerline method (Fig. 2-8):

$$17 \text{ ft} + 13 \text{ ft} + 17 \text{ ft} + 13 \text{ ft} = 60 \text{ ft}$$

Unit method (Fig. 2-9):

$$
\begin{array}{rcl}
2 \times 90 & = & 180 \text{ ft} \\
2 \times 33 & = & 60 \text{ ft} \\
2 \times 10 & = & \underline{20 \text{ ft}} \\
& & 260 \text{ ft}
\end{array}
$$

Perimeter method (Fig. 2-9):

90 ft + 35 ft + 30 ft + 10 ft + 30 ft + 10 ft + 30 ft + 35 ft

$$= 270 \text{ ft} - 4 \text{ ft} = 266 \text{ ft}$$

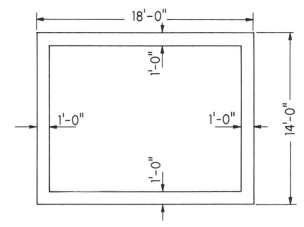

Figure 2–8

Center-line method (Fig. 2-9):

> 89 ft + 34 ft + 29 ft + 10 ft + 31 ft + 10 ft + 29 ft + 34 ft = 266 ft

Self-Testing Problems

Find the total stretch-out of walls for the plans shown in Fig. 2-10. Check your answers with those given in the Answer Box.

Answer Box	
Problem	Total Stretch-Out
1	48'0"
2	56'0"
3	118'8"
4	160'0"

MATERIAL AND INSTALLATION COSTS

> *Concrete:* Regular concrete weighs 150 lb per cubic foot; lightweight concrete weighs 90 to 110 lb per cubic foot. Concrete with the proper water–cement ratio can achieve a compressive strength of from 1000 to 6000 psi. Residential concrete has a compressive strength of 2000 to 3500 psi. The cost for concrete is $58.75 per ton.

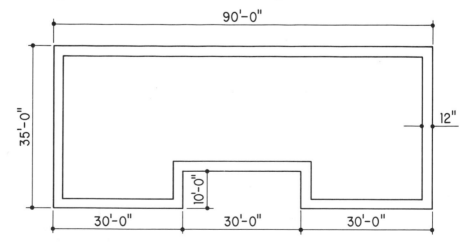

Figure 2–9

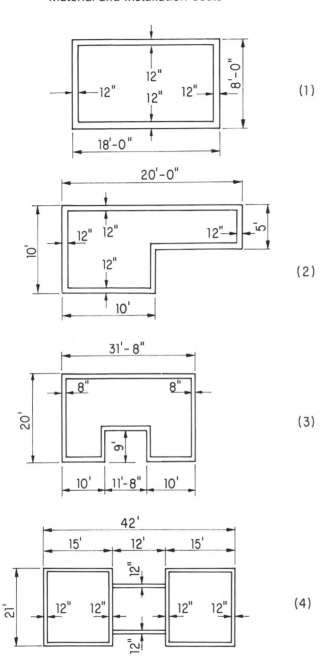

Figure 2-10

Ready-mix concrete: The cost is $44.95 per cubic yard.

Concrete ingredients:

Plain portland cement: $3.85 per bag

Sand washed for concrete: $7.35 per cubic yard

Stone $\frac{3}{4}$ to $1\frac{1}{2}''$: $9.90 per cubic yard

Concrete in place: Placing concrete for continuous footing, vibrating, labor, and equipment is $7.00 per cubic yard.   For forms and reinforcing steel, plus hoist equipment, materials, and labor the total cost is $175.70 per cubic yard.

SELF EXAMINATION

Find the cubic yards of concrete footing, and the cubic yards of concrete foundation wall for the plan shown in Fig. 2-11.   Use the unit method of taking-off the quantities.   Write your figures in the proper places in the tabular take-off form below or rule out your own take-off sheet on a separate sheet of paper.   Check your answers with those given in the Answer Box.

CONCRETE FOUNDATIONS—UNIT METHOD

Item	Unit	Length	Width	Depth	Cubic feet	Cubic yards
Footing	*1*	214.68	1.83	1.0	392.865	14.55
Walls	*1*	214.68	.83	7.3	1300.75	48.2

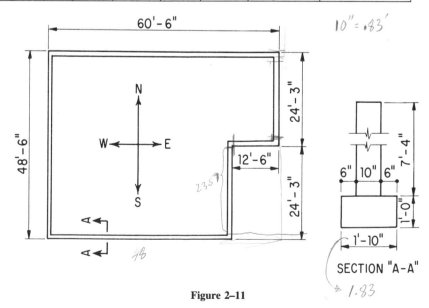

$10'' = .83'$

$23.5'$

48

$\star \ 1.83$

Figure 2–11

Answer Box	
Footings	14.5, or 15, cu yd
Walls	~~58 or 59~~ cu yd ⟵ *@ 1' WALL THICKNESS*

48 CY ⟵ @ 10" WALL THICKNESS

ASSIGNMENT—2

1. On an $8\frac{1}{2}$ in. × 11 in. white sheet of paper, sketch the following plan and section (Fig. 2-12) on the upper half of the sheet. On the lower half, prepare a tabular take-off of the quantities of concrete for the footings and foundation walls.

2. Design a 3000-psi concrete using a $1:2\frac{1}{4}:3$ mix. Find all aggregates.

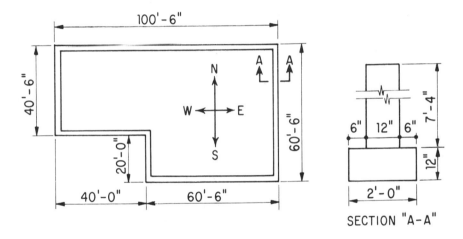

Figure 2–12

SECTION "A-A"

CONCRETE FOUNDATIONS

Item	Unit	Length	Width	Depth	Cubic Feet	Cubic Yards

Concrete foot
  North and south
  East and west
Walls
  North and south
  East and west						

SUPPLEMENTARY INFORMATION

QUANTITY OF CONCRETE FOR VARIOUS WALL THICKNESSES

Wall thickness (in.)	Cubic feet of concrete per square foot of wall	Cubic yards of concrete per square foot of wall	Wall thickness (in.)	Cubic feet of concrete per square foot of wall	Cubic yards of concrete per square foot of wall
3	0.25	0.0092	16	1.33	0.049
4	0.33	0.0122	18	1.5	0.055
6	0.50	0.0185	24	2.0	0.074
8	0.67	0.025	30	2.5	0.092
10	0.83	0.031	36	3.0	0.111
12	1.00	0.037	42	3.5	0.13
14	1.17	0.043	48	4.0	0.148

QUANTITY OF CONCRETE FOR VARIOUS SIZES OF FOOTINGS

Footing dimensions (in.) Height	Width	Cubic feet of concrete per foot of length	Cubic feet of concrete per 10 ft of length	Cubic yards of concrete per foot of length	Cubic yards of concrete per 10 ft of length
6	6	0.25	2.5	0.0092	0.092
6	8	0.33	3.33	0.0122	0.122
6	10	0.416	4.16	0.0154	0.154
8	8	0.436	4.36	0.0161	0.161
8	10	0.549	5.49	0.0203	0.203
8	12	0.667	6.67	0.0247	0.247
10	10	0.689	6.89	0.0255	0.255
10	12	0.834	8.34	0.0308	0.308
12	12	1.00	10.00	0.0370	0.370
12	14	1.167	11.67	0.0432	0.432
12	16	1.33	13.33	0.0492	0.492
12	18	1.5	15.00	0.0555	0.555
12	20	1.67	16.76	0.0618	0.618
12	22	1.83	18.34	0.0677	0.677
12	24	2.00	20.00	0.0740	0.740
14	24	2.33	23.33	0.0863	0.863
14	26	2.64	26.45	0.0977	0.977
16	26	2.88	28.82	0.1066	1.066
16	28	3.09	30.99	0.1144	1.144
18	30	3.75	37.50	0.1388	1.388

REVIEW QUESTIONS

1. Name the three methods used for taking-off concrete.
2. Five gallons of water per sack of cement makes a stronger concrete than 4 gal of water. True or false?
3. For a concrete mix of 1:3:5 using 6 gal of water, find the number of sacks of cement, the cubic feet of sand, and the cubic feet of gravel for a concrete slab 25 ft × 15 ft × 5 in.
4. Find the costs of ingredients used in Problem 3.

3

REINFORCINGS FOR FOUNDATIONS, PIERS, AND COLUMNS

INTRODUCTION

This unit concerns itself with estimating of reinforcing bars (rebars), used in footings, foundation walls, columns, piers, and concrete lintels. Rebars add strength to concrete and act also as a heat-transfer agent for temperature changes, thereby preventing cracking of the concrete. Rebars are deformed and are placed near the bottom of footings and slabs and near the edges of walls.

To understand the reason for the location of the bars, assume a board or plank supported at both ends. If a person were to stand on the board, it would bend under his or her weight. The degree of bending depends on the thickness of the board and the weight of the person. The wood fibers near the bottom of the board are stretched, or in tension, while the fibers on the top of the board are in compression.

Similarly, a concrete footing of a certain length, supporting a foundation wall, acts as the loaded board. The rebars, strong in tension, will not be stretched, preventing the bending action of the footing.

The same principle applies to a wall with vertical bars near its outside surfaces. Here the tensile strength of the bars prevents the loaded wall from buckling. Bars in walls are usually spaced 11 in. apart both vertically and horizontally.

Loads imposed on concrete floor slabs often require rebars or reinforcing mesh near the bottom of the slab. Slabs subjected to hydrostatic pressure (water pressure from under the slab) will require rebars near the top of the slab.

THE TAKE-OFF

Concrete footings often require reinforcing bars to give additional strength to the footing. These steel bars are figured by their actual length and their diameters. All bars of equal diameters are added up, and the total length is multiplied by the weight per foot to arrive at the total tonnage. An additional 5% is allowed for overlaps at corners, waste, and overlaps at splices of 30 diameters of the bar. A No. 4 bar ($\frac{1}{2}$ in.) is $30 \times \frac{1}{2} = 15$ in. Reinforcing bars are given by numbers ranging from 2 to 11. All numbers represent eighths.

For example, a No. 3 bar is $\frac{3}{8}$ in. in diameter; a No. 6 bar is $\frac{6}{8}$in. or $\frac{3}{4}$in. in diameter. The following table gives the standard sizes and weights of reinforcing bars:

STANDARD SIZES AND WEIGHTS OF CONCRETE REINFORCING BARS

Former bar designation (in.)	New bar designation (in.)	Unit weight (lb/ft)	Diameter (in.)	Cross section (sq in.)
$\frac{1}{4}$ Round	2	0.167	0.250	0.05
$\frac{3}{8}$ Round	3	0.376	0.375	0.11
$\frac{1}{2}$ Round	4	0.668	0.500	0.20
$\frac{5}{8}$ Round	5	1.043	0.625	0.31
$\frac{3}{4}$ Round	6	1.502	0.750	0.44
$\frac{7}{8}$ Round	7	2.044	0.875	0.60
1 Round	8	2.67	1.000	0.79
$1\frac{1}{8}$ Round	9	3.4	1.128	1.00
$1\frac{1}{4}$ Round	10	4.303	1.270	1.27
$1\frac{3}{8}$ Round	11	5.313	1.410	1.56

A Typical Problem

Find the total weight of reinforcings required in the footings of the plan shown in Fig. 3-1. The reinforcings are No. 5, or $\frac{5}{8}$ in. in diameter.

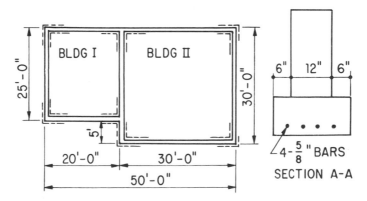

Figure 3–1

Solution   Reinforcing rods are allowed to overlap at corners.   Add all the wall dimensions and allow 5% extra, the amount extending beyond the faces of the wall in the footings and the intermediate overlaps and waste.   Therefore, considering all footings running east to west, 50 ft + 20 ft + 30 ft = 100 ft; considering all footings running north and south, 25 ft + 30 ft + 30 ft = 85 ft and 85 ft + 100 ft = 185 lin ft.   Since there are 4 bars in the footings, multiply the total length by 4:

$$4 \times 185 \text{ lin ft} = 740 \text{ lin ft}$$

$$\text{plus } 5\% = 740 \text{ lin ft} \times 1.05 = 777 \text{ lin ft}$$

$$777 \text{ lin ft} \times 1.043 \text{ lb/ft (weight per foot of } \tfrac{5}{8}\text{-in. bar)} = 810 \text{ lb}$$

$$810 \text{ lb} \div 2000 \text{ (2000 lb} = 1 \text{ ton)} = 0.405 \text{ or } 0.41 \text{ ton}$$

The reinforcing bars in the footings may be taken-off in tabular form in the following manner:

REINFORCING IN FOOTINGS ($\tfrac{5}{8}$-IN. BARS)

Item	Unit	Length	Total length	Weight per linear foot	Pounds	Tons
Footing under north and south walls	8	50	400			
Footing under west wall, bldg.	4	25	100			
Footing under east and west walls, bldg. II	8	30	240			
		Add 5%	740			
			37			
		Total	777	1.043	810	0.41

Estimating Reinforcings in Concrete Columns

Vertical reinforcing bars are placed near each corner of the column and extend from the footing to the top of the column (Fig. 3-2).   Lateral ties, spaced a specified distance, wrap around the vertical bars.   To create a better bond between the reinforced concrete and its footing, metal rods or dowels are allowed to protrude from the footing before the reinforced concrete column is cast.   On top of the column is the billet plate, which is mounted on a setting plate anchored to the column.   The billet plate receives the steel column, which is either bolted or welded to the billet plate.

The height of the concrete column is found by taking the difference between the minus dimensions of 4′0″ and 12′8″.   The minus 4′0″ means that the top of the

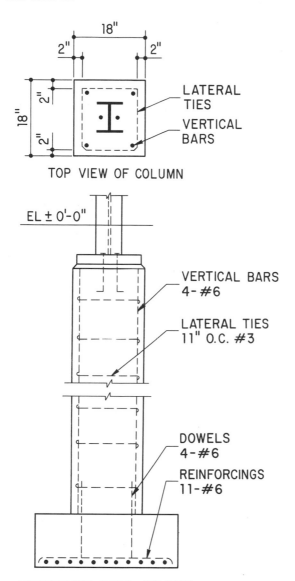

LATERAL
TIES

VERTICAL
BARS

TOP VIEW OF COLUMN

EL ± 0'-0"

VERTICAL BARS
4-#6

LATERAL TIES
11" O.C. #3

DOWELS
4-#6

REINFORCINGS
11-#6

REINFORCED CONC. COLUMN   Figure 3–2

concrete column is 4'0" below the finished first floor.   The bottom of the column
is 12'8" below the finished first floor.   The difference is

$$
\begin{array}{r}
12'8'' \\
-\ \ 4'0'' \\
\hline
8'8''
\end{array}
$$
    actual height of the column

Therefore,

$$\text{Vertical bars} \quad = 8.67 \text{ ft} \times 4 \quad = \quad 34.68 \text{ ft}$$

$$\text{Dowels} \quad = 2.67 \text{ ft} \times 4 \quad = \quad 10.68 \text{ ft}$$

$$\text{Footings} \quad = 3.5 \text{ ft} \times 22 \quad = \quad \underline{77 \text{ ft}}$$

$$122.36 \text{ lin ft}$$

122.36 lin ft $\times$ 1.502 lb/ft
(weight per foot of No. 6 rod; see table) = 183.78, or 184, lb

To figure the lateral ties, find the length of one tie. The column is 18 in. $\times$ 18 in. Since the lateral ties are wrapped around the vertical bars, each length is: 14 in. + 14 in. + 14 in. + 14 in. = 56 in. The laterals are spaced 11 in. on center. The column height is 8'8". Divide the height by the spacing to get the number of laterals, such as

$$8.67 \text{ ft} \div 0.91 \text{ ft} = 9.4, \text{ or } 10, \text{ laterals}$$

$$10 \times 56 \text{ in.} = 560 \text{ lin in.}$$

$$560 \text{ lin in.} = 46.7 \text{ lin ft} \times 0.376 \text{ lb/ft} = 17.56 \text{ lb}$$

Then the total weight is

$$\begin{array}{r} 184.00 \text{ lb} \\ + \underline{\ 17.56 \text{ lb}} \\ 201.56 \text{ lb} \ \div \ 2000 = 0.1007 \text{ ton} \end{array}$$

Allow 5% for cutting, overlap, and waste:

$$0.1007 \text{ ton} \times 1.05 = 0.1057 \text{ ton}$$

Estimating Reinforcings in Concrete Lintels

A lintel is a horizontal member of wood, steel, or concrete, placed over an opening to support the load above. From the plan, determine the number of interior openings requiring reinforcing concrete lintels. Find the width of each opening and allow an additional 8 in. for the bearing of the lintel (4 in. on either side of the openings unless otherwise noted). Multiply the length of the lintel (which is also the length of the reinforcing rod) by the number of reinforcings in the lintel to get the total linear feet. Multiply the linear feet by the number of openings to get the total linear feet of the rod. This is multiplied by the weight of the rod (see the table on page 00) to get the total pounds. Divide the pounds by 2000 to get the tonnage.

For example, take off the quantity of reinforcing rods required for the lintels over the openings shown on the plan given in Fig. 3-3. Note that lintels over

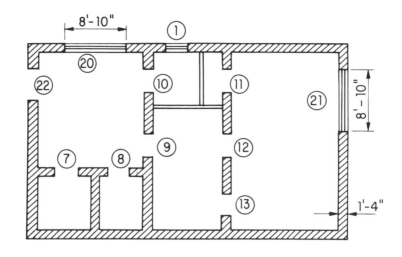

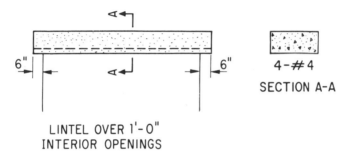

LINTEL OVER 1'-0"
INTERIOR OPENINGS

Figure 3–3

interior openings in 1'0" walls will be concrete reinforced with four pieces of $\frac{1}{2}$-in.-diameter rods. Lintels over exterior openings in 1'4" walls will be concrete reinforced with six pieces of $\frac{1}{2}$-in.-diameter rods (see the schedule for masonry openings).

SCHEDULE FOR MASONRY
OPENING DIMENSION

Opening number	Size of opening
1	3'0' × 7'0"
7	3'0" × 7'0"
8	3'0" × 7'0"
9	3'0" × 7'0"
10	2'6" × 7'0"
11	2'8" × 7'0"
12	3'0" × 7'0"
13	3'0" × 7'0"

Tabular Take-Off of Reinforcings

Following is the tabular take-off of reinforcings in concrete lintels used on the plan shown in Fig. 3-3.

Item	Number of openings	Bars per opening	Length of bar (ft)	Total length (ft)	Bar weight per foot	Pounds	Tons
Interior openings							
7, 8, 9, 12, 13	5	4	4	80.0			
10	1	4	3.5	14.0			
11	1	4	3.67	14.67			
Exterior openings							
20, 21	2	6	9.83	117.96			
22	1	6	5.33	31.88			
1	1	6	4.00	24.00			
				282.62			
			+5%	14.13			
				296.75	0.668	198.396	0.10

Problem

Estimate the tons of reinforcings in the footings shown in the plan and section in Fig. 3-4.

Solution

$$\text{East and west footing} = 30 \text{ ft} + 30 \text{ ft} = \quad 60 \text{ ft}$$

$$\text{North footing} = 100 \text{ ft}$$

$$\text{South footing} = 32 \text{ ft} + 32 \text{ ft} + 36 \text{ ft} = 100 \text{ ft}$$

$$\text{Returns (15 ft dim.)} = 15 \text{ ft} + 15 \text{ ft} = \underline{\quad 30 \text{ ft}}$$

$$290 \text{ lin ft}$$

$$3 \text{ bars} \times 290 = 870 \text{ lin ft}$$

Allow 5% for cutting, overlap, and waste:

$$870 \text{ ft} \times 1.05 = 913.5, \text{ or } 914 \text{ lin ft}$$

$$\text{Weight per foot} = 1.502 \text{ lb/ft} \times 914 \text{ lin ft} = 1373 \text{ lb}$$

$$1373 \text{ lb} \div 2000 = 0.69 \text{ ton}$$

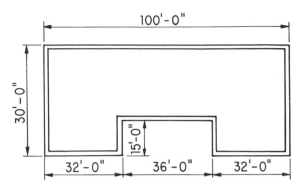

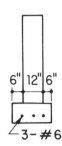

Figure 3–4

MATERIAL AND INSTALLATION COSTS

Reinforcings in footings: Four bar people can install 2.1 tons of No. 3 to No. 7 light bars at a cost of $855 per day, or 3.6 tons of No. 8 to No. 14 heavy bars at a cost of $655 per day. These costs inlude overhead and profit.

Bar chairs: For bar chairs up to 3 in. high, installed, the costs are:

$162 per thousand (M) plain

$202 per M galvanized

$335 per M stainless steel

$237 per M plastic

Reinforcings in columns and walls: Four bar people can install 1.6 tons of No. 3 to No. 7 light bars at a cost of $975 per day, or 2.7 tons of No. 8 to No. 14 heavy bars at a cost of $785 per day. These costs include overhead and profit.

Placing and tying of reinforcings: For footings, production runs from 9 hr per ton for heavy bars to 15 hr per ton for light bars. In beams, columns, and walls production runs from 12 hr per ton for heavy bars to 20 hr per ton for light bars. The overall average is about 14 hr per ton.

Tie wire: For each ton of bars, 5 lb of 16-gauge black annealed wire is required.

SELF EXAMINATION

Solve the following problem, and check your answer with that given in the Answer Box. If you have difficulty, review the text.

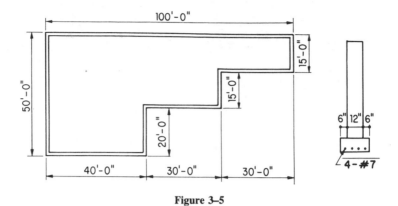

Figure 3–5

Estimate the tons of reinforcings shown in the plan and section of Fig. 3-5.

Answer Box
Total linear feet = 100
Plus 5% = 1260
Weight per foot = 2.044
Total pounds = 25.75
Total tons = 1.29

ASSIGNMENT—3

Sketch the plan and section given in Fig. 3-6 on the upper half of an $8\frac{1}{2}$ in. × 11 in. sheet of paper.   On the lower half of the paper, take-off in tabular form the total tonnage of steel reinforcings shown in the footings.   Bar sizes are No. 7.   Find the cost of bars and hours required for installation.

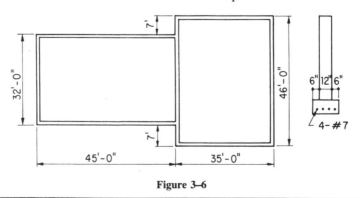

Figure 3–6

ASTM SPECIFICATIONS FOR REINFORCING BARS		TENSILE STRENGTH lb. per sq. in.	YIELD POINT lb. per sq. in.
A15	Intermediate Grade New Billet Steel	70,000–90,000	40,000 min.
A15	Hard Grade New Billet Steel	80,000 min.	50,000 min.
A16	Rail Steel Regular Grade	80,000 min.	50,000 min.
A61	Rail Steel Special Grade	90,000 min.	60,000 min.
A432	New Billet Steel	90,000 min.	60,000 min.
A431	New Billet Steel	100,000 min.	75,000 min.

REINFORCING BAR

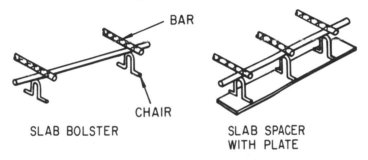

BAR

CHAIR

SLAB BOLSTER

SLAB SPACER WITH PLATE

Figure 3–7

SUPPLEMENTARY INFORMATION

Reinforcing bars in footings and slabs must be raised on bar chairs in order that the concrete may flow under and around the bars (Fig. 3-7). In a 12-in. thick concrete footing, the reinforcing bars are usually $1\frac{1}{2}$ to 3 in. from the bottom of the footing, depending on the moisture content of the soil.

REVIEW QUESTIONS

1. A No. 5 reinforcing bar is of what diameter and weighs how many pounds per linear foot?
2. Lateral ties in reinforced concrete columns are usually how far apart?
3. In estimating reinforcings in footings, what is the usual percentage allowance for overlap and waste?
4. What is the purpose of the setting plate on top of a reinforced concrete column?
5. What is the weight of steel per cubic inch?

4

FORMS FOR FOOTINGS, WALLS, AND TRENCHES

INTRODUCTION

In this unit you will learn how to estimate the quantity of wood formwork to receive the wet concrete of footings and foundation walls as well as forms used for sheet piling and bracing deep trench excavations.

Forms for concrete hold the concrete in place until it has hardened. Forms may be of wood, steel, or synthetic materials and they must be carefully erected to conform to the shape and dimensions of the concrete. The forms must also be sufficiently tight to prevent the leakage of water from the wet cement.

Consideration should be given to the cost of the formwork, because it constitutes a large part of the expense of concrete construction. Today, in most areas, building codes require that forms be designed by the structural engineer.

Spruce and pine lumber are well suited for wood forms, because they do not stain the exposed concrete surface. Frequently, wood forms are oiled or treated to fill the pores of the wood to prevent the absorption of water from the concrete and provide for a smoother surface, thus preventing the concrete from sticking to the forms when they are removed.

Footing forms may be constructed with 2 in. × 12 in. planks held in place by 2 in. × 2 in. stakes, 2 ft long, driven halfway into the ground. The stakes are held together by 1 in. × 3 in. space ties 4 ft apart.

Forms for concrete walls may be tongue-and-groove boards properly supported by ties, spreaders, stakes, and braces. Plywood form boards 1 ft × 8 ft, 2 ft × 8 ft, and 4 ft × 8 ft reinforced with 2 × 4's are also commonly used.

48

Major Requirements of Formwork

Formwork must be made strong enough to support the weight of the wet concrete and to overcome any sideward forces or pressures exerted by the concrete. Concrete weighs approximately 150 lb per cubic foot.

If a wall 6 ft high and 1 ft wide and 1 ft in length were to be considered, it would contain 6 cu ft of concrete, which has a total weight of 900 lb. This weight causes a considerable sideward force against the forms but is overcome by proper bracing and by the use of form ties, which extend through the concrete and are fastened to the other sides of the studs of the formwork.

THE TAKE-OFF

The amount of formwork required for concrete footings is determined by the amount of lumber touching wct concrete plus stakes, spreader ties, nails, and waste.

For example, estimate the forms required for the footing on the plan and section of Fig. 4-1.

To estimate the amount of formwork, find the total stretch-out of the footings and multiply this by 2 for both sides of the footing. Therefore, using the center-line method for the stretch-out:

$$59 \text{ ft} + 39 \text{ ft} + 59 \text{ ft} + 39 \text{ ft} = 196 \text{ ft} \times 2 \text{ (both sides)}$$

$$= 392 \text{ lin ft}$$

The linear feet are multiplied by the depth of the footing form to get the square feet of contact: 392 ft × 1 ft = 392 sq ft of form contact. Add from 5 to 10% for waste, according to ground conditions. For this problem let us assume a 7% waste; then: 392 ft × 1.07 ft = 419 sq ft. Since the wood items are estimated in board feet, let us first find out the exact meaning of a board foot.

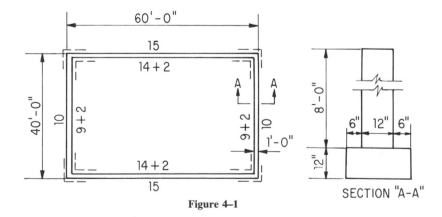

Figure 4–1

Board Foot or Board Measure—What Does It Mean?

On rough carpentry, lumber is figured according to the number of board feet. Simply stated, a board foot of lumber is equal to 144 cu in. of wood.   Assume for a moment that you have 144 cu in. of molding clay in your hands that can be worked into various shapes.

For example, the clay can be molded into a shape of 1 in. × 12 in. × 12 in. If you multiply these figures, the answer is 144 cu in., or 1 board foot (B.F.). Similarly, a piece of wood 4 in. × 6 in. × 6 in. when multiplied equals 144 cu in., or 1 B.F.   Figure 4-2 shows various shapes of wood which equal 144 cubic inches or 1 board foot.

Now suppose that we want to find the number of board feet of lumber in a board which measures 1 in. × 8 in. × 12 ft long.   The rule is to multiply the thickness by the width in inches by the length in feet, and divide the product by 12, such as

$$\frac{1 \text{ in.} \times 8 \text{ in.} \times \cancel{12} \text{ ft}}{\cancel{12}} = 8 \text{ B.F.}$$

For another example, a 2 in. × 4 in. stud, 8 ft long, contains how many board feet?

$$\frac{2 \text{ in.} \times \cancel{4} \text{ in.} \times 8 \text{ ft}}{\underset{3}{\cancel{12}}} = \frac{16}{3} = 5\tfrac{1}{3} \text{ B.F.}$$

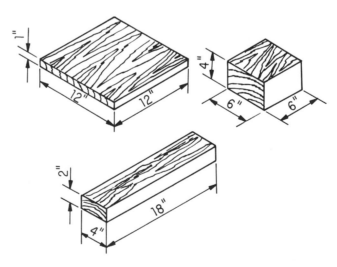

Figure 4–2

For a third example, a 4 in. × 4 in. post 10 ft long contains how many board feet?

$$\frac{4 \text{ in.} \times \cancel{4} \text{ in.} \times 10 \text{ ft}}{\underset{3}{\cancel{12}}} = \frac{40}{3} = 13.3 \text{ B.F.}$$

Now, getting back to Fig. 4-1, we found that the footings required a total of 392 lin ft. This plus a 7% waste equals 410 lin ft. Estimate the forms for the footings.

The form for the footing is comprised of the following items (Fig. 4-3):

1. Contact form off 2 in. × 12 in.
2. Stakes, 2 in. × 2 in. × 24 in. long spaced 4'0" on center (o.c.).
3. Spreader ties, 1 in. × 3 in. at 4'0" o.c.
4. Nails, $2\frac{1}{2}$ in. common—allow six nails per 4'0" o.c. of stakes and spreaders.

1. *Contact form*: The total linear feet of a 2 in. × 12 in. contact form, including a 7% waste, is equal to 419 lin ft. Then

$$\frac{419 \text{ lin ft} \times 2 \text{ in.} \times \cancel{12} \text{ in.}}{\cancel{12}} = 838 \text{ B.F.}$$

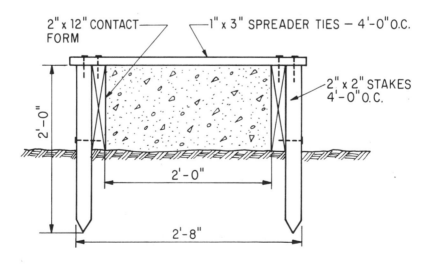

CONCRETE FOOTING FORM

Figure 4–3

Assume that the contact form members are in lengths of 12'0".

$$419 \div 12 = 35 \text{ pieces (pc.) approximately}$$

Therefore

$$\frac{35 \text{ pc.} \times 12 \text{ ft} \times 2 \text{ in.} \times 12 \text{ in.}}{12} = 840 \text{ B.F.}$$

2. *Stakes*: These are 2 in. × 2 in. × 24 in. long and are spaced 4'0" o.c. The perimeter or the stretch-out of the footing is 196 ft. Therefore,

$$\frac{196}{4} = 49 \text{ pairs of stakes}$$

49 pairs is 98 pc. 2 in. × 2 in. × 24 in. long; 98 pc. × 2 ft long = 196 lin ft of stakes. Allow 10 to 15% waste, according to ground conditions. Assume 12% waste for this problem.

$$196 \times 1.12 = 219.52 \quad \text{or} \quad 220 \text{ lin ft}$$

$$220 \text{ lin ft} \div 2'0" \text{ lengths} = 110 \text{ pc.}$$

$$\frac{110 \text{ pc.} \times 2\text{-ft lengths} \times 2 \text{ in.} \times 2 \text{ in.}}{12} = 73\tfrac{1}{3} \text{ B.F.}$$

3. *Spreader ties*: These are 1 in. × 3 in. at 4'0" o.c. and 2'8" long. The total stretch-out of footing is 196 ft.

$$\frac{196}{4} = 49 \text{ pc. of } 1" \times 3" \times 2'8" \text{ long}$$

The total linear feet of 1 in. b× 3 in. equals

$$49 \times 2.67(2'8") = 130.83 \text{ lin ft}$$

Assume 16% waste:

$$130.83 \times 1.16 = 151.7 \quad \text{say} \quad 152 \text{ lin ft}$$

Then

$$\frac{152 \text{ ft} \times 1 \text{ in.} \times 3 \text{ in}}{\underset{4}{12}} \overset{38}{=} 38 \text{ B.F.}$$

4. *Nails*: These are $2\frac{1}{2}$-in. common nails, six nails for each $4'0''$ o.c. of stakes.

$$\frac{196}{4} = 49 \times 6 \text{ nails} = 294 \text{ nails}$$

Allow 10% for waste, say, 324 nails for the job. Convert nails to pounds. The table in Fig. 4-4, gives the number of nails in 1 lb. There are 100 nails of $2\frac{1}{2}$-in. common in 1 lb. 324 nails is therefore

$$\frac{324}{100} = 3.24 \text{ lb}$$

of nails required.

The take-off of the formwork for the footings in Fig. 4-1 may now be expressed in tabular form.

NUMBER OF NAILS TO THE POUND		
Size	Length	Common
2 d	$1''$	847
3 d	$1\frac{1}{4}''$	543
4 d	$1\frac{1}{2}''$	296
5 d	$1\frac{3}{4}''$	254
6 d	$2''$	167
7 d	$2\frac{1}{4}''$	150
8 d	$2\frac{1}{2}''$	101
9 d	$2\frac{3}{4}''$	92.1
10 d	$3''$	66
12 d	$3\frac{1}{4}''$	66.1
16 d	$3\frac{1}{2}''$	47
20 d	$4''$	30
30 d	$4\frac{1}{2}''$	23
40 d	$5''$	17.3
50 d	$5\frac{1}{2}''$	13.5
60 d	$6''$	10.7

Figure 4-4

FORMS FOR FOOTINGS

Item	Unit	Length	Total length	Waste (%)	Total length plus % waste	B.F.
Footing: 2 in. × 12 in. contact forms	2	196	392	1.07	419	840
Stakes: 2 in. × 2 in. by 2'0" long	98	2	196	1.12	220	73⅓
Spreader: ties 1 × 3 × 2'8" at 4'0" O.C.	49	2.67	130.83	1.165	152	38

Nails: 2½-in. common

6 nails for each 4'0" o.c.

$$\frac{196}{4} = 49 \times 6 = 294 \text{ nails}$$

Plus 10% waste = 324 nails = 3.24 lb

Concrete Formwork—Basement Walls

The plywood forms are the most common forms in use today. These are usually made up by the contractor and are semipermanent. The number of times that such forms may be used depends on how they are handled when placing and dismantling. Some contractors will get as few as 5 uses, whereas others may be able to use them as many as 50 times.

Plywood Forms

These are usually made up in three sizes, although other sizes do exist. The plywood is of ¾ in., framed with 2 in. × 4 in. dimension lumber (Fig. 4-5). Each form has a top and bottom plate and is reinforced with sheet metal strapping on all outside corners of the form. The 2 × 4's are secured with 3½-in. common nails, while the plywood is fastened to the frame with 2½-in. common nails. The stud spacing for the 4 ft × 8 ft plywood form is 16 in.; for the 2 ft by 8 ft form and the 1 ft × 8 ft form, 12 in.

Estimating the plywood forms. Plywood forms may be estimated by the number of forms of each size required for the job.

For example, estimate the number of plywood forms required for the foun-

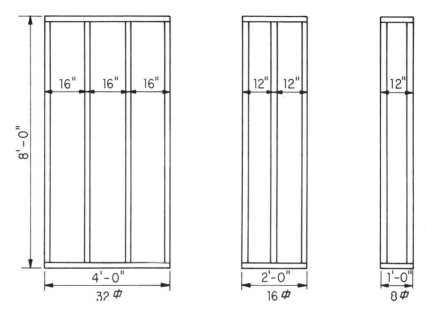

Figure 4–5

dation wall in Fig. 4-6. Assume 4 ft × 8 ft plywood forms for the 60-ft north and south exterior wall lengths.

The number of forms required are

$$60 \div 4 = 15 \text{ forms (north exterior face)}$$

$$= 15 \text{ forms (south exterior face)}$$

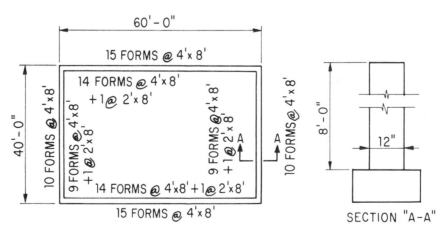

Figure 4–6

The interior faces of the north and south walls are 58 ft long.

$$58 \div 4 = 14 \text{ forms at } 4 \text{ ft} \times 8 \text{ ft plus } 1 \text{ form at } 2 \text{ ft} \times 8 \text{ ft}$$
$$= 14 \text{ forms at } 4 \text{ ft} \times 8 \text{ ft plus } 1 \text{ form at } 2 \text{ ft} \times 8 \text{ ft}$$

The 40-ft east and west exterior wall lengths are similarly divided:

$$40 \div 4 = 10 \text{ forms (east exterior wall)}$$
$$= 10 \text{ forms (west exterior wall)}$$

The interior faces of the east and west walls are 38 ft long:

$$38 \div 4 = 9 \text{ forms at } 4 \text{ ft} \times 8 \text{ ft plus } 1 \text{ form at } 2 \text{ ft} \times 8 \text{ ft}$$
$$= 9 \text{ forms at } 4 \text{ ft} \times 8 \text{ ft plus } 1 \text{ form at } 2 \text{ ft} \times 8 \text{ ft}$$

Add all 4 ft × 8 ft panels.

$$15 + 15 + 14 + 14 + 10 + 10 + 9 + 9 = 96 \text{ panels}$$

Summarize:

$$96 \text{ plywood forms at } 4 \text{ ft} \times 8 \text{ ft} = 32 \text{ sq ft} \times 96 = 3072 \text{ sq ft}$$
$$4 \text{ plywood forms at } 2 \text{ ft} \times 8 \text{ ft} = 16 \text{ sq ft} \times 4 = 64 \text{ sq ft}$$

In the example above we required

$$96 \text{ plywood panels } (4 \text{ ft} \times 8 \text{ ft} \times \tfrac{3}{4} \text{ in.})$$
$$\text{plus } 4 \text{ plywood panels } (2 \text{ ft} \times 8 \text{ ft} \times \tfrac{3}{4} \text{ in.})$$

Each 4 ft × 8 ft panel has 40 lin ft of 2 in. × 4 in.

Each 2 ft × 8 ft panel has 28 lin ft of 2 in. × 4 in.

$$
\begin{aligned}
40 \text{ ft} \times 96 \text{ ft} &= 3840 \text{ lin ft of } 2 \times 4\text{'s} \\
28 \text{ ft} \times 4 \text{ ft} &= \underline{112 \text{ lin ft of } 2 \times 4\text{'s}} \\
&= 3952 \text{ lin ft of } 2 \times 4\text{'s}
\end{aligned}
$$

Practically no waste allowance is needed.   The board feet are

$$\frac{3952 \times 2 \times 4}{\frac{12}{3}} = 2634.66, \text{ say } 2635, \text{ B.F.}$$

Nail requirements.   10 lb of $3\frac{1}{2}$-in. common nails per 1000 B.F. for the 2 by 4's.   Therefore, $2.635 \times 10 = 26.35$ lb of nails.   Add 10% for waste = 29 lb required.

20 lb of $2\frac{1}{2}$-in. common nails per 1000 sq ft of plywood form.   Total square feet of plywood = 3136 sq ft. $3.136 \times 20 = 62.72$ lb of nails.   Add 10% for waste = 69 lb required.

Galvanized sheet requirements. 20 sq ft of galvanized sheet cut for strapping the corners of the frames per 1000 sq ft of form. Then $3.136 \times 20 = 62.72$ sq ft of sheet. It is generally agreed that it takes 20 hr of carpenter time per 1000 sq ft of completed form and 5 hr of carpenter helper's time for 1000 sq ft of completed form. In Fig. 3-6 the total contact area of the form is

$$
\begin{array}{ll}
60 \text{ ft} + 60 \text{ ft} + 40 \text{ ft} + 40 \text{ ft} = & 200 \text{ lin ft} \\
58 \text{ ft} + 58 \text{ ft} + 38 \text{ ft} + 38 \text{ ft} = & \underline{192 \text{ lin ft}} \\
& 392 \text{ lin ft}
\end{array}
$$

Then 392 lin ft $\times$ 8 ft (height of wall) = 3136 sq ft of form. Carpenter's time = 3.136×20 hr = 62.72 hr. Helper's time = 3.136×5 hr = 15.68 hr. Local wage rates should be multiplied by the hours above to arrive at the cost of labor for building the forms for this job.

SELF EXAMINATION

Now you have the opportunity to find out how well you have learned the material thus far. For the plan and section shown (Fig. 4-7), estimate the following. Check your answers with those given in the Answer Box.

Footing Forms

1. 2 in. $\times$ 10 in. contact forms (allow 7% waste).
2. 2 in. $\times$ 3 in. stakes, 18 in. long, 4'0" o.c. (allow 15% waste).
3. 1 in. $\times$ 3 in. spreaders, 2'10" long, 4'0" o.c. (allow 16% waste).
4. $2\frac{1}{2}$-in. common nails, six nails per 4'0" o.c. (allow 10% waste).

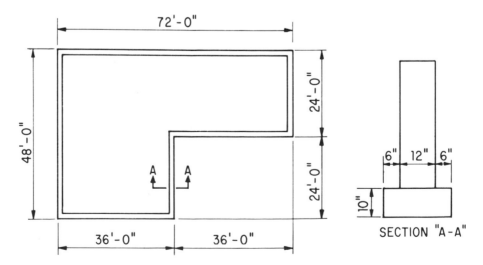

Figure 4-7

Answer Box	
2 in. b × 10 in. contact form 7% waste included	841.7 B.F.
2 in. × 3 in. stakes, 18 in. long at 4'0" o.c. 15% waste included	102 B.F.
1 in. × 3 in. spreaders, 2'10" long at 4'0" o.c. 16% waste included	48.5 B.F.
2½-in. common nails six nails per 4'0" o.c. 10% waste included	3.89, or 4, lb

Figuring Bracing and Sheet Piling for Trenches

Sheet piling is required where the soil is not self-supporting or where excavations adjoin a property line.   This is estimated by the square foot, by taking the number of square feet of bank or trench walls to be braced or sheet piled and estimating the cost of the work at a certain price per square foot for labor and lumber required.

Let us determine the quantity of lumber needed to sheet pile a trench on both sides, 60 ft long and 8 ft deep (Fig. 4-8).   The total area to be sheet piled is

$$60 \text{ ft} \times 8 \text{ ft} = 480 \text{ sq ft}$$

$$480 \times 2 = 960 \text{ sq ft for both sides}$$

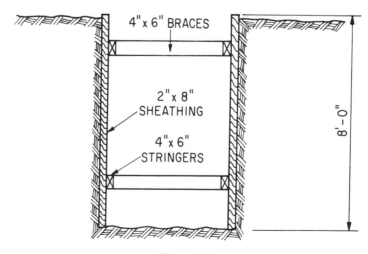

Figure 4–8

Dressed 2 in. × 8 in. sheathing is $1\frac{1}{2}$-in. × $7\frac{1}{4}$ in. × 8 ft long. Dressed 4 in. × 6 in. stringers are $3\frac{1}{2}$ in. × $5\frac{1}{2}$ in. × 12 ft long. Dressed 4 in. × 6 in. braces are $3\frac{1}{2}$ in. × $5\frac{1}{2}$ in. × 3 ft long, spaced 5 ft apart.

Board feet of sheathing planks.   When planks are placed side by side for a length of 60 ft, then 60 ÷ 0.60 ($7\frac{1}{4}$ in.) = 100 planks.

$$100 \times 2 = 200 \text{ planks for both sides}$$

One plank has

$$\frac{200 \text{ pc.} \times 1\frac{1}{2} \text{ in.} \times 7\frac{1}{4} \text{ in.} \times 8 \text{ ft}}{12} = 1450 \text{ B.F.}$$

Board feet of stringers.   Figure 4-8 shows four stringers, $3\frac{1}{2}$ in. × $5\frac{1}{2}$ in. × 60 ft trench length.

$$\frac{4 \text{ pc.} \times 3\frac{1}{2} \text{ in.} \times 5\frac{1}{2} \text{ in.} \times 60 \text{ ft}}{12} = 385 \text{ B.F.}$$

Board feet of braces.   The braces are $3\frac{1}{2}$ in. × $5\frac{1}{2}$ in. × 3 ft long, spaced 5'0" apart. In a 60-ft length there will be 60 ÷ 5 = 12 spaces plus 1, or 13 pc., 3 ft long × 2 rows = 26 pc. total length of braces is 26 × 3 = 78 lin ft.

$$\frac{3\frac{1}{2} \text{ in.} \times 5\frac{1}{2} \text{ in.} \times 78 \text{ ft}}{12} = 125 \text{ B.F.}$$

Deep excavations.   Sheet piling for deep excavations is shown in Fig. 4-9. Assume that the earth wall to be supported is 48 ft and 8 ft high. Find the board feet of the various pieces of lumber.

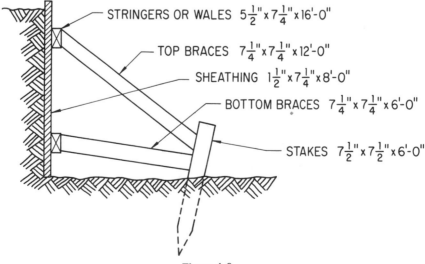

STRINGERS OR WALES $5\frac{1}{2}$"x$7\frac{1}{4}$"x16'-0"

TOP BRACES $7\frac{1}{4}$"x$7\frac{1}{4}$"x12'-0"

SHEATHING $1\frac{1}{2}$"x$7\frac{1}{4}$"x8'-0"

BOTTOM BRACES $7\frac{1}{4}$"x$7\frac{1}{4}$"x6'-0"

STAKES $7\frac{1}{2}$"x$7\frac{1}{2}$"x6'-0"

Figure 4-9

Sheathing. $1\frac{1}{2}$ in. $\times$ $7\frac{1}{4}$ in. $\times$ 8 ft planks are used.   The trench length is 48 ft. Then:

$$48 \div 60 \ (7\frac{1}{4} \ \text{in.}) = 80 \ \text{pc. of sheathing}$$

$$80 \ \text{planks} \times 7.25 \ \text{B.F. in one plank} = 580 \ \text{B.F.}$$

Stringers or Wales. $5\frac{1}{2}$ in. $\times$ $7\frac{1}{4}$ in. $\times$ 16-ft lengths are used.   The trench length is 48 ft.   Two stringers make 96 lin ft.   The number of 16-ft lengths is:

$$96 \div 16 = 6 \ \text{pc.}$$

The board feet are:

$$\frac{5\frac{1}{2} \ \text{in.} \times 7\frac{1}{4} \ \text{in.} \times 96 \ \text{ft}}{12} = 319$$

Top Braces. $7\frac{1}{4}$ in. $\times$ $7\frac{1}{4}$ in. $\times$ 12-ft lengths are used spaced 4'0" apart.   The length of the trench is 48 ft. Then:

$$48 \div 4 = 12 \ \text{spaces plus} \ 1 = 13 \ \text{pc.}$$

$$13 \ \text{pc. of} \ 12 \ \text{ft} = 13 \times 12 = 156 \ \text{lin ft}$$

The board feet are:

$$\frac{7\frac{1}{4} \ \text{in.} \times 7\frac{1}{4} \ \text{in.} \times 156 \ \text{ft}}{12} = 683 \ \text{B.F.}$$

Bottom Braces. $7\frac{1}{4}$ in. $\times$ $7\frac{1}{4}$ in. $\times$ 6 ft lengths spaced 4'0" apart, 13 pc. are required (same number as for top braces).

$$13 \ \text{pc.} \times 6 \ \text{ft} = 78 \ \text{lin ft}$$

$$\frac{7\frac{1}{2} \ \text{in.} \times 7\frac{1}{2} \ \text{in.} \times 78 \ \text{ft}}{12} = 683 \ \text{B.F.}$$

Stakes. $7\frac{1}{2}$ in. $\times$ $7\frac{1}{2}$ in. $\times$ 6 ft in length, spaced 4'0" apart. 13 pc. of 6 ft = 78 lin ft

$$\frac{7\frac{1}{2} \ \text{in.} \times 7\frac{1}{2} \ \text{in.} \times 78 \ \text{ft}}{12} = 366 \ \text{B.F.}$$

MATERIAL AND INSTALLATION COSTS

Every contractor's office has fixed costs which cannot easily be charged to any one project.   Such fixed costs are office rent, salaries of office personnel, telephone, stationery, insurance, and other items.   Total costs for these fixed overhead expenses may be calculated on a yearly basis and an hourly or weekly amount charged to each project that takes place throughout the year.

Every contracting firm as well as any other business is entitled to a profit. The markup for overhead and profit in retail stores is from 35 to 50% over wholesale costs. A contractor's markup of 30 to 40% is not unusual.

Wall forms: To install plywood forms for concrete walls up to 8 ft in height will cost $3.92 per square foot of contact area. This includes material and installation plus a 34% markup for overhead and profit.

Forms for columns: Columns 12 in. × 12 in. can be formed with plywood at a cost of $4.95 per square foot of contact area. This includes material and installation costs plus a 35% markup for the contractor's overhead and profit.

Forms for footings: To install contact forms for footings, including the necessary stakes, spreader ties, and contact planks, costs $2.49 per square foot of contact form. This includes material, installation costs, and a 33% markup for overhead and profit.

Matt foundation forms: To install forms for a matt foundation costs $5.05 per square foot of contact area. This includes material and installation costs plus a 37% markup for the contractor's overhead and profit.

SELF EXAMINATION

Check your answers with those given in the Answer Box.

1. A concrete wall 120 ft long and 8 ft high requires how many plywood form panels of what size?
2. 1 B.F. contains how many cubic inches?
3. 26 pc. of 2 in. × 12 in. × 12-ft lengths is how many board feet?
4. How many pounds of $2\frac{1}{2}$-in. common nails are required for 6975 sq ft of plywood forms? Include a 10% waste factor.
5. How many $3\frac{1}{2}$-in. common nails are required for 4165 B.F. of 2 in. by 4 in. form studs?
6. A 4 ft × 8 ft × $\frac{3}{4}$ in. plywood panel form requires how many board feet of form studs?
7. A 2 ft × 8 ft × $\frac{3}{4}$ in. plywood form panel requires how many linear feet of 2 in. by 4 in?
8. What lengths of 2 in. × 4 in. would you order to make a 1'0" × 8'0" plywood form panel?

Answer Box	
Problem	
1	60 forms of 4 ft × 8 ft
2	144 cu in.
3	624 B.F.
4	153.4 lb
5	41.65 lb
6	$26\frac{2}{3}$ B.F.
7	28 lin ft
8	One 2 in. × 4 in. at 8 ft long Two 9-ft lengths

ASSIGNMENT-4

Take-off the quantities of form work for the footing and foundation walls shown in Fig. 4-10.   Find the following:
Footings:

1. 2 in. × 12 in. contact form
2. 2 in. × 2 in. stakes, 4'0" o.c. by 24 in. long
3. 1 in. × 3 in. spreaders, 4'0" o.c.
4. $2\frac{1}{2}$-in. common nails
5. Material and installation costs, including overhead and profit.

Walls:

6. Number of plywood forms of each size.   Use panel sizes: 4'0" by 8'0", 2'0" by 8'0", and 1'0" by 8'0".
7. Total B.F. of 2 in. × 4 in. form studs.
8. Nail requirements: $2\frac{1}{2}$-in. common and
            $3\frac{1}{2}$-in. common.
9. Galvanized sheet for strapping.

10. Carpenter hours for completed forms.   Carpenter helper's hours for completed forms.

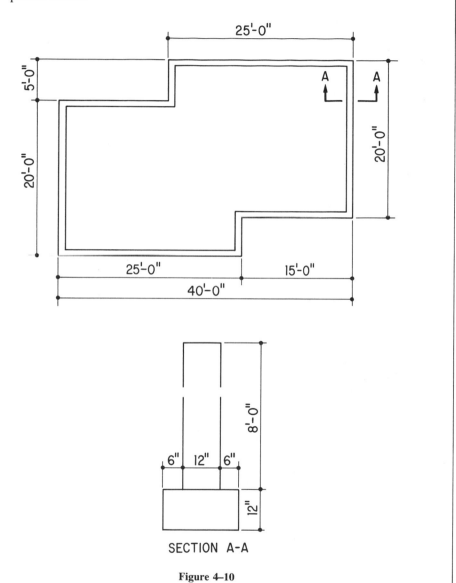

SECTION A-A

Figure 4–10

SUPPLEMENTARY INFORMATION

DATA ON SHEET PILING

Size of timber		B.F.M. single sheet	Number of pieces for:			
Cross section	Length		25 ft	50 ft	75 ft	100 ft
2 in. × 8 in. square edged	7'0"	9.3	38	75	113	150
3 in. × 8 in. square edged	8'0'	16	38	75	113	150
2 in. × 10 in. square edged	9'0"	15	30	60	90	120
3 in. × 10 in. square edged	12'0"	30	30	60	90	120
3 in. × 12 in. square edged	14'0"	42	25	50	75	100

Forms for Beams and Girders

To determine the number of square feet of formwork for reinforced beams and girders, add the three sides of the beam and girder and multiply by the length. The dimensions of the beam or girder are taken from the underside of the slab to the bottom of the beam and girder.

For example, in Fig. 4-11, add 9 in. + 9 in. + 9 in. = 27 in., or 2'3", or 2.25 ft.

$$2.25 \text{ ft} \times \text{length of beam} = \text{sq ft of forms}$$

It is considered good practice to estimate formwork for beams and girders separately. Forms are generally of 1-in. material. Where a smooth surface is desired, plywood should be used in place of sheathing.

Estimating Forms for Concrete Slab

The form boards shown in Fig. 4-12 are of 1-in. lumber supported by 2 in. × 8 in. joists 8 ft long spaced 24 in. o.c. The ends of the joists are supported by 2 in. × 6 in. vertical members also spaced 24 in. o.c. The ribbons of 1 in. × 6 in. material further support the joists. Assuming a bay length of 20'0", the following quantities of lumber are required:

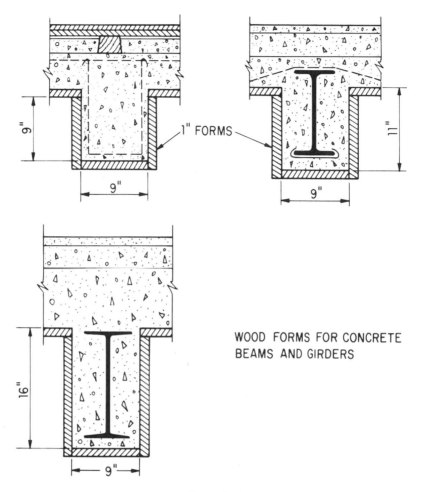

1" FORMS

WOOD FORMS FOR CONCRETE
BEAMS AND GIRDERS

Figure 4–11

Area of forms (supporting slab) 20 ft × 8 ft = 160 sq ft plus
 15% waste   = 184 B.F.
Joists, 20 ft ÷ 2 ft = 10 joists plus 1 = 11 joists − 11
 pieces at 2 in. × 8 in. × 8 ft   = 118 B.F.
Ribbons, 2 ft × 20 ft = 40 lin ft of 1 in. × 6 in.   = 20 B.F.
Vertical 2 in. × 6 in. members, 22 pieces 2′3″ long, or
 2.25 ft × 22   = 50 B.F.
                                                                  372 B.F.

Since 1 bay has 160 sq ft, then 372 ÷ 160 = 2.31 B.F. per square foot of surface.
Note: Forms for beams should be figured separately.

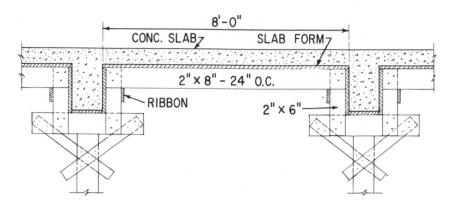

Figure 4–12

REVIEW QUESTIONS

1. A board foot of lumber contains how many cubic inches?
2. A 2 in. × 4 in. × 8 ft stud is how many board feet?
3. 35 pieces of 2 in. × 12 in. × 24 ft members contain how many board feet?
4. List the wooden members required for a footing form.
5. A ten-penny nail is how long? How many nails to the pound?
6. What is a contractor's percentage markup?
7. Find the cost of formwork for an ordinary concrete footing 12 in. deep by 24 in. wide and a stretch-out length of 140 ft. Include a 38% markup for overhead and profit.

5

CONCRETE FLOORS, FILL, AND CEMENT FINISH

INTRODUCTION

This unit covers the quantity take-off of concrete for floors, together with stone fill, cement finish, concrete steps, sidewalks, and other work. Reinforced concrete is used in residential as well as in large constructions for footings, foundations, slabs on grade, steps, walks, and innumerable other areas where the loads are great enough for rebar or mesh reinforcement.

Concrete is called *reinforced concrete* when rebars or reinforcing mesh is installed. Mesh or welded-wire fabric is reinforcement that uses wires of the same or different diameter, welded together to form a grid. Such reinforcement not only strengthens the slab but acts as a heat-transfer agent for temperature changes.

On drawings mesh is designated by gauge numbers for ferrous and nonferrous types; it is also designated by diameter in decimals of an inch, as shown in the table of "Wire Gauges of Ferrous and Nonferrous Wire" in the Supplementary Information section of this unit.

Steel wire mesh reinforcing for concrete is indicated as follows:

$$WWF6 \times 6-W1.4 \times W1.4$$

Where WWF refers to woven-wire fabric, the first 6 represents the longitudinal wire spacing, and the second 6 represents the transverse wire spacing. The W1.4 × W1.4 represents the longitudinal wire size and the transverse wire size, both are 10 gauge. (see the table on wire gauges). The symbols are used on both structural and architectural drawings.

Similarly, WWF8 × 12–W7 × W4 indicates that longitudinal wires are 8 in. on center and transverse wires are 12 in. on center. Longitudinal wires are 0 gauge with the transverse wires 4 gauge (again, see the table of wire gauges).

Crushed stone fill is often used under concrete slabs for leveling before the slab is poured, and where wet conditions prevail under the slab. Crushed stone not larger than $1\frac{1}{2}$ in. is suitable for many types of construction and is also used as aggregate in concrete.

Lightweight concrete is characterized by lightweight aggregate, which is also generally used as fill because it is light in weight, has a good insulating value, has fire-resisting properties, and is nailable. Finished surfaces of concrete may be smooth, grooved, rough, polished, broomed, or given various types of pebble and stone chip finishes.

The maximum size of aggregate is governed by the nature of the concrete work. In thin slabs or walls, the largest pieces of aggregate (crushed stone) should not exceed one-fifth to one-fourth of the thickness of the section of concrete being placed. In reinforced concrete, specifications usually limit the maximum size to three-fourths of the minimum clear spacing between rebars.

THE TAKE-OFF

Concrete floor slabs are estimated by multiplying the length of the slab in feet by the width of the slab in feet by the thickness of the slab in feet to get the cubic feet. The cubic feet are divided by 27 to get the cubic yards.

For example, a concrete slab measuring 16 ft × 16 ft × 4 in. in thickness contains the following number of cubic yards of concrete:

$$16 \text{ ft} \times 16 \text{ ft} \times 0.33 \text{ ft} = 84.48 \text{ cu ft}$$

$$84.48 \text{ cu ft} \div 27 = 3.13 \text{ cu yd}$$

A Typical Problem

Find the cubic yards of concrete required for the cellar floor shown in Fig. 5-1.

Solution   To begin, divide the plan into rectangular portions, such as indicated by the diagonal lines, creating areas I and II. The thickness of the slab is 4 in. Section AA (Fig. 5-1) shows the concrete slab against the inner face of the foundation wall. The quantities in tabular form are taken off in the following manner:

Item	Unit	Length	Width	Depth	Cubic feet	Cubic yards
Area I	1	26	22	0.33	188.76	
Area II	1	34	14	0.33	157.08	
					345.84	12.8

Note: Dimensions given on the plan are exterior dimensions. To get the slab dimensions it is necessary to subtract the thickness of the wall, which is 12 in., from each dimension at both ends. The 26-ft dimension, however, has not changed because at the right end of the dimension 1 ft is subtracted, while at the left end of the dimension 1 ft is added. On the 36-ft dimension, 1 ft is deducted from each end, making it 34 ft.

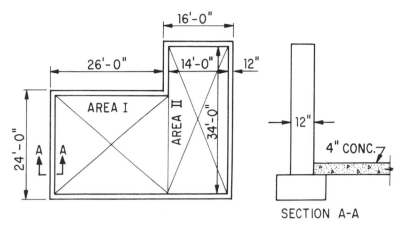

Figure 5–1

Stone Fill Take-Off

The quantity of stone fill is found by multiplying the length by the width by the depth of the fill in feet to get the cubic feet. Divide the cubic feet by 27 to get the cubic yards.

A Typical Problem

Find the number of cubic yards of stone fill required under the concrete slab shown in the plan and section of Fig. 5-2.

Solution Divide the plan into three rectangles as indicated by the areas marked I, II, and III. Note that the diagonals are up to the hidden lines that represent the footing lines. The section indicates that the stone bed is against the footings but not against the foundation wall.

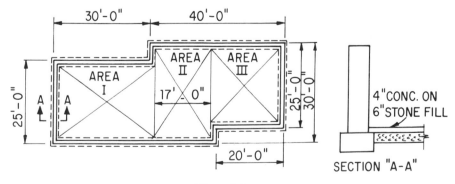

Figure 5–2

The simplest procedure and perhaps the best method is to take-off the quantities in tabular form. It is important to check the following figures against the plan in Fig. 5-2:

STONE FILL UNDER FLOOR

Item	Unit	Length	Width	Depth	Cubic feet	Cubic yards
Area I	1	30	22	0.5	330.0	
Area II	1	27	17	0.5	229.5	
Area III	1	22	20	0.5	220.0	
					779.5	28.87

Lightweight Concrete

Lightweight concrete is estimated similarly to stone fill. Multiply the length of the slab by the width by the depth to find the cubic feet. Divide the total cubic feet by 27 to find the cubic yards. Lightweight concrete is made in much the same manner as ordinary concrete except that cinders or other lightweight aggregates are used in place of the stone aggregate. Cinder concrete can be used for fireproofing and for fill over structural floor and roof slabs. The cinders should be well burned, vitreous, clinker, and reasonably free of sulfides. Sulfides occurring in considerable quantity cause reinforcing steel bars to rust. Cinder concrete may also be used on roof slabs to create the desired pitch for the drainage of water.

A Typical Problem

Find the cubic yards of lightweight concrete fill used on the roof in Fig. 5-3.

Solution The roof fill is 6 in. at its high point and 2 in. at its low point. This makes an average depth of 4 in. Following is the take-off of the lightweight fill in tabular form:

LIGHTWEIGHT CONCRETE ROOF FILL

Item	Unit	Length	Width	Depth	Cubic feet	Cubic yards
Roof deck	1	58.67	28.67	0.33	555.06	21

The figure 58.67 in the tabular take-off is found by subtracting twice the thickness of the wall from the 60-ft length. Since the wall is 8 in., twice the thickness is 1'4".
The 28.67 dimension is found similarly by subtracting 1'4", twice The thickness of the wall, from the 30-ft dimension.

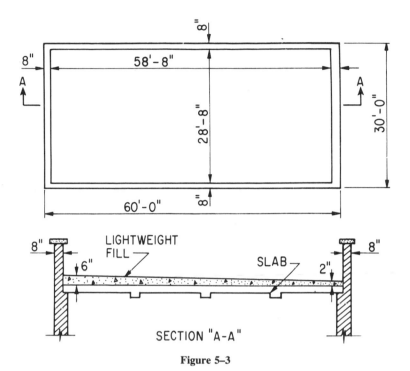

Figure 5–3

Estimating Concrete Steps

Multiply the thickness of the slab under the trends by the length of the slab by the width of the slab in feet to get the cubic feet of the slab. Next, find the cubic feet of each triangle.

For example, Fig. 5-4 shows a typical concrete stair. The slab thickness is 6 in. The length of the slab is 6 ft. The width of the slab is 4 ft. Then

$$0.5 \text{ ft} \times 6 \text{ ft} \times 4 \text{ ft} = 12 \text{ cu ft}$$

Next, the tread width is 10 in. or 0.83 ft, the riser height is 7 in. or 0.58 ft, the tread length is 4 ft. Therefore,

$$0.83 \text{ ft} \times 0.58 \text{ ft} \times 4 \text{ ft} = 1.926 \text{ cu ft} \div 2 = 0.963 \text{ cu ft}$$

for one triangle. Then

$$0.963 \text{ cu ft} \times 6 \text{ ft (for six triangles)} = 5.778, \text{ or } 5.8, \text{ cu ft}$$

Total 12 cu ft

+ 5.8 cu ft

17.8, or 18 cu ft of concrete, or $\frac{2}{3}$ cu yd

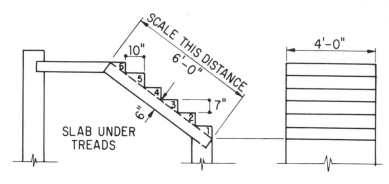

Figure 5–4

Estimating Concrete for Sidewalks

Find the number of square feet of surface by multiplying the length times the width in feet. Walks of irregular shape can be broken up into squares, rectangles, and so on; the area figured for each is added together.

The table in Fig. 5-5 shows the amounts of cement, sand, and stone necessary for 100 sq ft of concrete walk.

MATERIALS REQUIRED FOR 100 SQUARE FEET OF SIDEWALKS AND FLOORS FOR UNVARYING THICKNESSES									
Proportions	1 : 1½ : 3			1 : 2 : 3			1 : 2 : 4		
Thickness in inches	Cement in Cu. ft.	Sand in Cu. yds.	Stone in Cu. yds.	Cement in Cu. ft.	Sand in Cu. yds.	Stone in Cu. yds.	Cement in Cu. ft.	Sand in Cu. yds.	Stone in Cu. yds.
2½	5.9	0.33	0.65	5.4	0.40	0.60	4.6	0.34	0.68
3	7.0	0.39	0.78	0.65	0.48	0.72	5.6	0.41	0.82
3½	8.2	0.46	0.91	7.5	0.56	0.84	6.5	0.48	0.96
4	9.4	0.52	1.04	8.6	0.64	0.95	7.4	0.55	1.10
5	11.7	0.65	1.3	10.8	0.80	1.19	9.3	0.69	1.37

Figure 5–5

For example, the sidewalks shown in Fig. 5-6 are divided into sections marked I, II, III, IV, and V.   Multiply the length and width of each section, and add the areas.   Thus

Area I = 5 ft × 5 ft = 25 sq ft

Area II = 15 ft × 4 ft = 60 sq ft

Area III = 40 ft × 4 ft = 160 sq ft

Area IV = Here you must find the length of the centerline of the curved sidewalk. This is found the same way as a circumference of a circle. Since the length of the centerline is only one-fourth of the circle, the circumference is divided by 4.

The formula for the circumference of a circle is

$$C = \pi \times D \quad \text{or} \quad C = 3.14 \times 12 \text{ ft} = 37.68 \text{ ft} \div 4 = 9.42 \text{ ft}$$

Then

$$9.42 \text{ ft} \times 4 \text{ ft} = 37.68 \text{ sq ft}$$

$$\text{Area V} = 18 \text{ ft} \times 4 \text{ ft} = 72 \text{ sq ft}$$

$$\text{Total square feet} = \quad 25$$

$$60$$

$$160$$

$$37.68$$

$$72$$

$$354.68 \text{ sq ft or } 355 \text{ sq ft}$$

Assuming that a mix of 1:2:3 is used and the sidewalks have a thickness of 4 in., find the sacks of cement, the cubic yards of sand, and the cubic yards of stone.

Referring to Fig. 5-5, under the column having a 1:2:3 proportion, and opposite the thickness of the slab, you will find the following:

  8.6 sacks of cement
 17.28 cu ft of sand
 25.65 cu ft of gravel

These quantities are for 100 sq ft of surface.   Since our computations show a total of 355 sq ft of surface, it is necessary to multiply each of the figures above by 3.55. Therefore,

$$8.6 \text{ sacks} \times 3.55 = 30.53, \text{ or } 31, \text{ sacks of cement}$$

$$17.28 \text{ cu ft} \times 3.55 = 61.34 \text{ cu ft of sand}$$

$$25.65 \text{ cu ft} \times 3.55 = 91 \text{ cu ft of gravel}$$

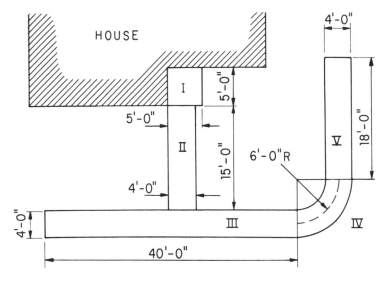

Figure 5–6

Estimating Concrete in a Beam and Slab Floor

Multiply the length by the width by the thickness of the slab in feet to find the cubic feet. The quantity of concrete required for the concrete beams is found by multiplying the height by the width by the length of the beam in feet to find the cubic feet. Multiply this by the number of beams and divide the total cubic feet by 27 to find the total cubic yards.

For example, estimate the cubic yards of concrete required for the beam and slab construction in Fig. 5-7. Take-off the quantities in tabular form.

TAKE-OFF FOR FIG. 5-7

Item	Unit	Length	Width	Depth	Cubic feet	Cubic yards
Beam slab	1	40	16	0.33	211.2	
Beams	4	16	0.75	0.75	36	
					247.2	9.1

MATERIAL AND INSTALLATION COSTS

Welded-wire fabric—WWF6 × 6–No. 10/10–(W1.4 × W1.4): The material and installation cost of 100 sq ft of this welded-wire fabric is $23, which includes a 34% contractor's markup for overhead and profit.

Welded-wire fabric—WWF2 × 2: No. 14 galvanized wire for beam and

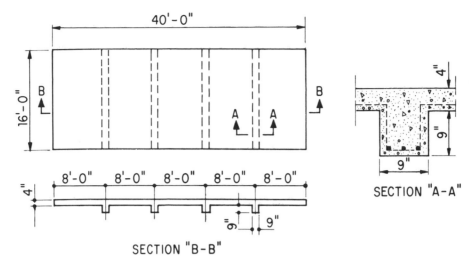

Figure 5–7

column wrap costs $91 per 100 sq ft of material and installation. The price includes a 45% contractor's markup for overhead and profit.

Waterproof membrane on concrete slab: For 1-ply felt, material and installation cost is $0.64 per square foot. Add a percentage for overhead and profit.

Elastomeric waterproofing: Plain $\frac{1}{32}$-in.-thick neoprene sheets cost $0.96 per square foot for material and installation. A contractor's markup is not included.

Sidewalks: Concrete (3000 psi) with 6 × 6–No. 10 mesh, broomed finish, no base, 4 in. thick costs $1.88 per square foot in place. This includes a 25% markup for contractor's overhead and profit.

Concrete paving with mesh: Using 4500-psi concrete, 6-in. thick, costs $12.20 per square yard for material and installation. An 8-in.-thick slab costs $15.25 per square yard for material and installation. This does not include a contractor's markup for overhead and profit.

Crushed stone: Use $\frac{3}{4}$ in. in size, used as the base course at $7.00 per ton when placed 3 in. deep, will cost $1.31 per square yard; 6 in. deep will cost $2.37 per square yard. This includes a contractor's markup of about 13% for overhead and profit.

Expansion joint: Bituminous fiber $\frac{1}{2}$ in. × 6 in. costs $0.82 per linear foot installed.

Vapor barrier: 0.006 thick, lapped 6 in., costs $5.45 per square for material and $4.15 per square for installation.

Rigid insulation: Flat sheets of fiberglass 8 oz per square foot cost $2.25 per square foot installed.

Concrete channel slabs: 3½-in.-thick slabs cost $1.92 per square foot, and $0.75 per square foot for installation.

Lightweight ready-mix concrete: A 1:4 mix over structural roof slab costs $81.45 per cubic yard.

SELF EXAMINATION

From the plan and section in Fig. 5-8, take-off the following quantities: (1) stone concrete cellar floor slab, (2) lightweight concrete over slab, and (3) stone fill under slab.

STONE CONCRETE, LIGHTWEIGHT CONCRETE, STONE FILL

Item	Unit	Length	Width	Depth	Cubic feet	Cubic yards
4-in. Stone concrete						
Area 1						
Area 2						
3-in. lightweight concrete						
Area 1						
Area 2						
6-in. stone fill						
Area 1						
Area 2						

Answer Box	
4-in. stone concrete	16.91, or 17, cu yd
3-in. lightweight concrete	12.81, or 13, cu yd
6-in. stone fill	24 cu yd

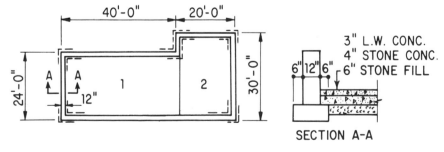

Figure 5–8

ASSIGNMENT—5

On a sheet of paper $8\frac{1}{2}$ in. × 11 in., sketch on the upper half of the sheet the plan and section given in Fig. 5-9. On the lower half, in tabular form take-off the quantities of materials of the following:

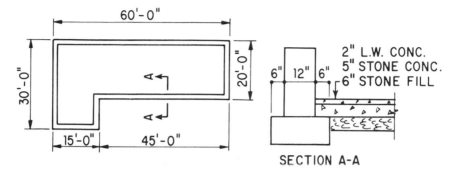

Figure 5–9

(1) a 5-in. stone concrete floor using a $1:1\frac{1}{2}:3$ mix; (2) a 6-in. stone fill under floor; and (3) a 2-in. lightweight concrete floor slab. Find the quantities of cement, sand, and gravel for the 5-in. concrete floor.

SUPPLEMENTARY INFORMATION

Recommended Thicknesses of Concrete Slabs (Inches)

Basement floors for dwellings	4
Private garage floors	4–5
Porch floors	4–5
Stock barn floors	5–6
Poultry house floors	4
Hog house floors	4
Milkhouse floors	4
Granary floors	5
Implement shed floors	6
Tile floor bases	$2\frac{1}{2}$
Driveways and approaches	6–8
Sidewalks	4–6

COLORS TO BE USED IN CONCRETE FLOOR FINISH

Color desired	Commercial names of colors for use in cement	Pounds of color required per sack of cement to obtain:	
		Light shade	Medium shade
Grays, blue-black, and black	Germantown lampblack	$\frac{1}{2}$	1
	or carbon black or	$\frac{1}{2}$	1
	black oxide of manganese	1	2
	or mineral black[a]	1	2
Blue shade	Ultramarine blue	5	9
Brownish red to dull brick red	Red oxide of iron	5	9
Bright red to vermillion	Mineral turkey red	5	9
Red sandstone to purplish red	Indian red	5	9
Brown to reddish brown	Metallic brown (oxide)	5	9
Buff, colonial tint, and yellow	Yellow ocher or yellow	5	9
	oxide	2	4
Green shade	Chromium oxide or	5	9
	greenish-blue ultramarine	6	

[a]Only first-quality lampblack should be used. Carbon black is light and requires very thorough mixing. Black oxide or mineral black is probably most advantageous for general use. For black, use 11 lb of oxide per sack of cement.

SQUARE FEET COVERAGE OF VARIOUS QUANTITIES OF CONCRETE AT VARIOUS THICKNESSES

Cubic yards of concrete	Thickness							
	1 in.	$1\frac{1}{4}$ in.	$1\frac{1}{2}$ in.	2 in.	$2\frac{1}{2}$ in.	3 in.	$3\frac{1}{2}$ in.	4 in.
1	324	216	185	162	130	108	93	81
$1\frac{1}{2}$	486	324	278	243	194	162	138	121.5
2	648	432	361	324	259	216	185	162
$2\frac{1}{2}$	810	540	454	405	324	270	231	202.5
3	972	648	547	486	389	324	277	243
$3\frac{1}{2}$	1134	756	640	567	454	378	323	283.5
4	1296	864	733	648	519	432	369	324
$4\frac{1}{2}$	1458	972	826	729	584	486	415	364.5
5	1620	1080	919	810	649	540	461	405
$5\frac{1}{2}$	1780	1188	1012	891	714	594	507	445.5
6	1942	1296	1105	972	779	648	553	486
$6\frac{1}{2}$	2164	1404	1198	1053	844	702	599	526.5

MATERIAL REQUIRED FOR 100 SQ FT OF CONCRETE FLOOR BASE

Thick-ness (in.)	Proportions								
	1:2:3			1:2:4			1:2½:5		
	Cement (bbl)	Sand (cu yd)	Pebbles (cu yd)	Cement (bbl)	Sand (cu yd)	Pebbles (cu yd)	Cement (bbl)	Sand (cu yd)	Pebbles (cu yd)
3	1.62	0.48	0.71	1.38	0.41	0.82	1.15	0.43	0.85
3½	1.89	0.56	0.83	1.61	0.48	0.96	1.35	0.50	1.00
4	2.16	0.64	0.95	1.84	0.55	1.10	1.54	0.56	1.23
4½	2.43	0.72	1.07	2.07	0.62	1.24	1.73	0.63	1.26
5	2.68	0.80	1.19	2.31	0.69	1.37	1.92	0.70	1.41

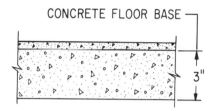

CONCRETE FLOOR BASE — 3"

To determine the quantity of cement (barrels), sand (cubic yards), and pebbles (cubic yards) in a concrete floor base measuring 10'0" × 15'0", proceed as follows:

$$10 \times 15 = 150 \text{ sq ft}$$

For 100 sq ft of 3-in. floor base, in the table above the quantities are:

$$\text{Cement} = 1.62 \text{ bbl}$$

$$\text{Sand} = 0.48 \text{ cu yd}$$

$$\text{Pebbles} = 0.71 \text{ cu yd}$$

Therefore, multiply

$$1.5 \times 1.62 = 2.43 \text{ bbl of cement}$$

$$1.5 \times 0.48 = 0.72 \text{ cu yd of sand}$$

$$1.5 \times 0.71 = 1.06 \text{ cu yd of pebbles}$$

For 225 sq ft of floor base, multiply the values in the table by 2.25.

AMOUNTS OF MATERIALS FOR 100 SQ FT OF WALL AREA

Thickness of wall (in.)	Proportions					
	1:2½:5			1:2:4		
	Cement (bbl)	Sand (cu yd)	Pebbles (cu yd)	Cement (bbl)	Sand (cu yd)	Pebbles (cu yd)
6	2.30	0.85	1.70	2.70	0.83	1.66
8	3.08	1.13	2.26	3.70	1.10	2.20
10	3.85	1.41	2.82	4.63	1.37	2.70
12	4.60	1.70	3.40	5.56	1.66	3.30
15	5.76	2.12	4.24	6.93	2.06	4.12
18	6.90	2.55	5.10	8.34	2.49	4.98

QUANTITIES FOR 100 SQ FT WEARING SURFACE OR TOPCOAT

Thickness (in.)	Proportions				
	1:2		1:1:1		
	Cement (bbl.)	Sand (cu yd)	Cement (bbl.)	Sand (cu yd)	Pebbles (cu yd)
½	0.51	0.15			
¾	0.75	0.23			
1	1.00	0.29	1.00	0.15	0.15
1¼	1.26	0.37	1.26	0.19	0.19
1½	1.51	0.45	1.51	0.23	0.23
2	2.00	0.59	2.00	0.30	0.30

SQUARE FEET OF SURFACE-COVERING CAPACITY OF MORTAR FROM ONE BAG OF CEMENT

Mixture of parts by volume		Thickness of coat				
Cement	Sand	¼ in.	⅜ in.	½ in.	¾ in.	1 in.
1	1	66	44	33	22	16
1	1½	84	56	42	28	21
1	2	101	67	50	33	25
1	2½	118	78	59	39	29
1	3	136	90	68	45	34
1	3½	153	102	76	51	42
1	4	171	113	85	57	38

Note: One bag of cement contains 94 lb net.

A barrel contains 376 lb net.

One barrel = 376 ÷ 94 = 4 bags of cement.

WIRE GAUGES OF FERROUS AND NONFERROUS WIRE

Gauge number	Ferrous American Steel Wire Gauge (in.)	Nonferrous American Wire Gauge (in.)	Woven-wire fabric-symbols
00	0.3310	0.3648	W8.5, W8, W7.5
0	0.3065	0.3249	W7, W6.5
1	0.2830	0.2893	W6, W5.5
2	0.2625	0.2576	W5
3	0.2437	0.2294	W4.5
4	0.2253	0.2043	W4, W3.5
5	0.2070	0.1819	W4, W3
6	0.1920	0.1620	W2.9
7	0.1770	0.1443	W2.5
8	0.1620	0.1285	W2.1, W2
9	0.1483	0.1144	W1.5
10	0.1350	0.1019	W1.4
11	0.1205	0.0907	
12	0.1055	0.0808	
13	0.1915	0.0720	
14	9.0800	0.0641	
15	0.0720	0.0571	
16	0.0625	0.0508	
17	0.0540	0.0453	
18	0.0475	0.0403	
19	0.0410	0.0359	
20	0.0348	0.0320	

REVIEW QUESTIONS

1. Explain the purpose of reinforcing bars and wire mesh in a concrete slab.
2. A 6-in. concrete slab 56 ft $\times$ 80 ft requires a one-ply felt waterproof membrane. How much material is required? What is the total cost?
3. What is the cost of constructing a sidewalk 4'0" wide $\times$ 150' long using 3000-psi concrete, 4 in. thick, with 6 $\times$ 6 No. 10 mesh?
4. In Problem 3, a bed of 6-in.-deep crushed stone of $\frac{3}{4}$-in. size is used under the concrete sidewalk. Find the total cost, including a contractor's markup of 14% for overhead and profit.
5. How many cubic yards of concrete are required to cover 378 sq ft of 3-in-thick concrete?

6

BRICK WORK

INTRODUCTION

Common and face brick are manufactured in standard sizes with minor variations in dimensions. All bricks vary slightly in their dimensions because of the shrinkage of the clays in burning. Most brick today is made with round or rectangular holes instead of being solid because the voids allow the brick to be evenly burned throughout in the kiln.

Brick may be named or identified by the particular placement within the wall. Ordinarily, as one views a brick wall, the bricks laid horizontally are called *stretchers*. When the brick lengths are placed at right angles to the length of the wall, the bricks are called *headers*. Bricks placed standing on end are known as *soldiers*.

Brick bonds are used primarily to assist in strengthening the brick wall. By bonding brick, various patterns or designs are created. In the common header bond, a header course is placed every fifth, sixth, or seventh course. The all-stretcher band is reinforced every sixth course with metal ties placed between the joints, and the stacked bond has similar metal reinforcing every sixth course.

Mortars for brick are available ready-mixed (in bags), premixed (mixed at plant and transported to site), and job-mixed (ingredients are mixed by hand or in a cement mixer at the site).

Mortars for large construction are controlled by laboratory testing to meet the requirements known as *property specifications*, or by predetermined proportions by volume, known as *proportion specifications*.   Sand for mortar is generally white and is graded by the size of the joint.

For small jobs there exist premixed mortars that can be obtained usually from a local building supplier.   For the small portable mixer the following mortar is used: 1:2:4, or 1 part portland cement to 2 parts lime putty to 4 parts of fine sand.

This unit deals with the most accurate method of estimating the quantity of brick required for a particular job, together with the amount of mortar for the brick joints.

THE TAKE-OFF

Accurate Brick Estimating

In estimating brick the size of the brick and the size of the mortar joint must be considered.   Mortar joints vary from $\frac{1}{8}$ in. to 1 in., with $\frac{1}{2}$ in. as the common joint. Specifications on larger jobs usually give the mortar joint that is to be used.   When this is known, and also the exact size of the brick is known, the following method is the most accurate:

For example, a nominal brick size measures 2 in. × 4 in. × 8 in. The actual size of the bricks, however, is $2\frac{1}{8}$ in. × $3\frac{3}{4}$ in. × $7\frac{1}{2}$ in. If we assume a bed mortar joint and an end mortar joint of $\frac{1}{2}$ in., the brick face with its mortar joints will measure $2\frac{5}{8}$ in. high by 8 in. long (Fig. 6-1).

If we multiply the total length of the brick by the total height of the brick (including mortar joints), we will get the square-inch surface.   Therefore, $2\frac{5}{8}$ in. × 8 in., or 2.625 in. × 8 in. = 21 sq in. (*Note*: $\frac{5}{8}$ in. expressed as a decimal is 0.625 in.)   We know that 1 sq ft is equal to 144 sq in. If one brick with its mortar joints takes up 21 sq in., 144 sq in. will take up

$$144 \text{ sq in.} \div 21 \text{ sq in.} = 6.86 \text{ bricks for a 4-in. wall}$$

If the wall is 8 in., or two bricks thick, multiply 2 × 6.86 = 13.72 bricks for an 8-in. wall.   If the wall is three bricks thick, multiply 3 × 6.86 = 20.58 bricks for a 12-in. wall.

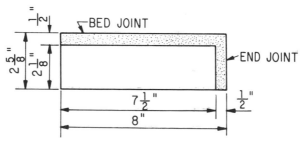

Figure 6-1

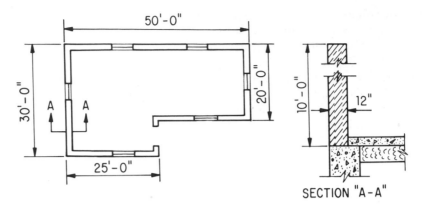

Figure 6–2

You can readily see that in using this method it is necessary to find the number of square feet of brick wall for the entire job. Deduct the square feet for windows, doors, and other openings and then multiply the net area of the wall by the number of bricks found for 1 sq ft of wall surface, bearing in mind the thickness of the wall.

Let us work out an actual problem. *For example*, find the number of common bricks in the plan and section given in Fig. 6-2. We will assume a brick size of $3\frac{3}{4}$ in. $\times$ 5 in. $\times$ 8 in. (a double-size common), where the actual height of the brick is 5 in. and the actual length is 8 in. (The width of the brick need not be considered.) We will also assume a bed and end mortar joint of $\frac{1}{2}$ in. (see Fig. 6-3). *Note*: The windows are $3'0'' \times 4'6''$; the door is $3'0'' \times 7'0''$. Therefore, the face area of the brick including the mortar joints is

$$5.5 \text{ in.} \times 8.5 \text{ in.} = 46.75 \text{ sq in.}$$

We know that 1 sq ft of brick surface is 144 sq in. If 1 brick with its mortar joints

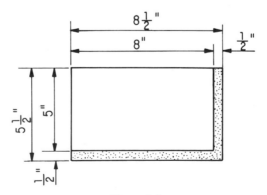

Figure 6–3

takes up 46.75 sq in., 144 sq in. will take up

$$144 \text{ sq in.} \div 46.75 \text{ sq in.} = 3.1 \text{ bricks for a 4-in. wall}$$

Since the wall is 12 in., or 3 bricks thick, we multiply the number of bricks for 1 sq ft by 3, which is

$$3 \times 3.1 = 9.3 \text{ bricks in 1 sq ft of wall}$$

Next, take-off the net square feet of wall in tabular form and multiply the total net square feet by 9.3 bricks. Look at the following tabular take-off, and check the figures by referring to the plan and section (Fig. 6-2).

COMMON BRICK (DOUBLE SIZE, $3\frac{3}{4}$ IN. × 5 IN. × 8 IN.)

Item	Unit	Length	Height	Square feet	Square feet out	Total
North and south wall	2	50.0	10.0	1000.0		
East and west wall	2	28.0	10.0	560.0		
Windows	6	3.0	4.5		81.0	
Door	1	3.0	7.0		21.0	
				1560.0	102.0	
				− 102.0		
				1458.0 × 9.3 =		13,560
				Plus 2%		× 1.02
						13,832

For your convenience, the number of bricks per square foot for various mortar joints and various wall thicknesses are given in Fig. 6-4; the brick size is the double-size common 8 in. × 5 in. × $3\frac{3}{4}$ in.

NUMBER OF DOUBLE SIZE COMMON BRICK FOR 1 SQ FT OF BRICK WALL (BRICK SIZE = $3\frac{3}{4}$ in. x 5 in. x 8 in.)

Thickness of Wall (in.)	Number of Bricks Thick	Mortar Joints (Bed and End)			
		$\frac{1}{4}$ in.	$\frac{3}{8}$ in.	$\frac{1}{2}$ in.	$\frac{5}{8}$ in.
4	1	3.32	3.2	3.08	3.17
8	2	6.64	6.4	6.16	6.34
12	3	9.96	9.6	9.3	9.51

Figure 6-4

Mortar Requirements

To find the cubic feet of mortar, divide the total number of bricks by 1000 and multiply by the factor given in the following table:

CUBIC FEET OF MORTAR REQUIRED FOR 1000 BRICKS (NO ALLOWANCE FOR WASTE INCLUDED; ADD 10% FOR WASTE)

Joint thickness (in.)	Wall thickness (in.)				
	4	8	12	16	20
$\frac{1}{4}$	5.7	8.7	9.7	10.2	10.5
$\frac{3}{8}$	8.7	11.8	12.9	13.4	13.7
$\frac{1}{2}$	11.7	15.0	16.2	16.8	17.1
$\frac{5}{8}$	14.8	18.3	19.5	20.1	20.5

For example, we have seen in the previous problem that 13,832 bricks are required. To find the total cubic feet of mortar, divide

$$13,832 \div 1000 = 13.832 \text{ thousands of bricks}$$

Since the wall is 12 in. thick and the mortar joints are $\frac{1}{2}$ in., multiply the factor 16.2 (found in the table) by the thousands of bricks, such as

$$13.832 \times 16.2 = 224 \text{ cu ft of mortar}$$

224 cu ft + 22.4 cu ft (10% for waste) = 246.4, say 247 cu ft of mortar required

MATERIAL AND INSTALLATION COSTS

Common brick: Prices vary in different parts of the United States. An average cost per thousand (M) is about $192. The installation cost per M brick is about $424, making a total of $618 per M. To this figure an overhead and profit allowance should be added—let us say 10%. This brings the total cost for material, installation, overhead, and profit to $679 per M.

Face brick: Prices are somewhat higher, where the average cost for material and installation is $757 per M. Including a 10% allowance for overhead and profit, the total cost is $842 per M. To the above must be added about $51 per M for mortar and scaffolding.

Brick veneer: select, common 4 in. thick, 8 in. $\times$ $2\frac{2}{3}$ in. $\times$ 4 in. at $206 per M. There are 6.75 brick in 1 sq ft. The cost for material including mortar and scaffolding ($51 per M) is $263 per M. The installation cost is $514 per M, or a total of $777 per M. Allow a markup for overhead and profit, say 30%.

Cleaning brick: A solution of 1 gal of muriatic acid to 20 gal of water is used in cleaning 1000 sq ft of smooth brick. 70 sq ft an hour can be cleaned by one person. On rough brick only 50 sq ft per hour can be cleaned.

SELF EXAMINATION

Find the total number of brick required for the plan and section shown in Fig. 6-5. On a sheet of paper $8\frac{1}{2}$ in. × 11 in. sketch the plan and section on the upper half of the sheet. On the lower half, in tabular form (Fig. 6-6), find the total number of bricks required. Check your answer with that given in the Answer Box. *Note*: The windows are 2'6" × 4'0"; the door is 3'0" × 7'0".

Answer Box	
Total square feet of wall	1,755
Number of bricks per square foot of 8-in. wall	× 12.33
	21,639
Allow 2% for waste	× 1.02
	22,072

Estimating Face Brick

Assume that the walls of the plan in Fig. 6-5 are laid in common bond with the exterior faces of standard face brick backed up with standard common brick. The face brick is bonded to the common brick with a full-header course every fifth course. We can assume that half of the total brick found is face and the other half common brick.

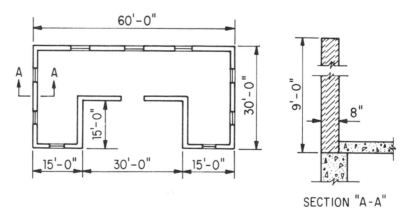

Figure 6–5

| BRICK SIZE : STANDARD AMERICAN $2\frac{1}{4}$" x $3\frac{3}{4}$" x 8" | | | | | | |
| MORTAR JOINT – 1/2 of an inch | | | | | | |
Item	U	L	H	Sq. ft.	Sq. ft. Out	Total Brick
N and S wall						
E and W wall						
Returns						
Windows						
Door						

Figure 6–6

This, of course, is not entirely true. The full headers of face brick every fifth course displace an amount of common brick, and therefore an additional allowance must be made for face brick; this same allowance must be deducted from the total common brick. Figure 6-7 gives the percentages that must be added to the face brick for the type of bond specified.

The total brick without waste (see the Answer Box) is 21,639. One-half of

PERCENTAGES ADDED TO FACE BRICK FOR VARIOUS BONDS		
Type of bond	Full headers	Percentages to be added
Common	Every 5th course	20% or $\frac{1}{5}$
Common	Every 6th course	16.7% or $\frac{1}{6}$
Common	Every 7th course	14.3% or $\frac{1}{7}$
English	Every 6th course	16.7% or $\frac{1}{6}$
English	Every other course	50% or $\frac{1}{2}$
Flemish	Every 6th course	5.6% or $\frac{1}{18}$
Flemish	Every course	33.3% or $\frac{1}{3}$

Figure 6–7

this is 10,819.5 face bricks plus 20% for a bond of every fifth course.   Therefore,

20% of 10,819.5 is 2164 bricks

10,819.5 + 2164 = 12,983.5 face bricks

Add 2% for waste = 12,983.5 × 1.02

= 13,243

From the common bricks, which total 10,819.5, subtract the percentage of face brick (20%) that displace the common brick.   Thus

10,819.5 − 2164 = 8656 common bricks

Add 2% for waste = 8656 × 1.02

= 8829 total common bricks

ASSIGNMENT—6

1. On $8\frac{1}{2}$ in. × 11 in. paper, prepare a quantity take-off in tabular form of the following materials for the plan and section in Fig. 6-8:
   (a) concrete footings;
   (b) concrete walls;
   (c) concrete slab;

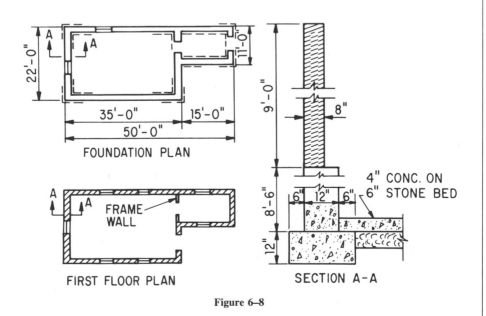

FOUNDATION PLAN

FIRST FLOOR PLAN

SECTION A-A

Figure 6–8

(d) 6-in. stone bed; and

(e) American Standard brick (6.33 per sq ft)

Windows, cellar: 2 ft × 3 ft

Door, cellar: 3 ft × 7 ft

Windows, first floor: 2'6″ × 6'6″

Door, first floor: 3 ft × 7 ft

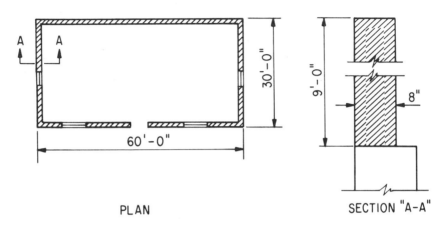

PLAN   SECTION "A-A"

Figure 6–9

2. Take-off in tabular form the quantity of common brick required for the plan and section shown in Fig. 6-9. The brick size is 8 in. long and $2\frac{1}{4}$ in. high. The mortar joint is $\frac{1}{2}$ in.

Windows: 2'6″ × 4'3″ and 6'0″ × 4'3″

Door: 3'0″ × 7'0″

3. For Problem 2, find the material and installation costs as well as the cost including overhead and profit.

Item	Unit	Length	Height	Square feet	Square feet out	Total brick
North and south wall						
East and west wall						
Windows						
Door						

SUPPLEMENTARY INFORMATION

Important Data on Various Types of Brick

Building brick. There are three classifications or grades of brick based on their resistance to weather conditions.

Grade SW: This brick is used where a very high degree of frost is encountered or where the brick may be frozen when permeated with water. These brick may be used for foundation courses and retaining walls.

Grade MW: This brick is used where freezing temperatures occur but is unlikely to be permeated with water. This brick may be used on the face of a wall above grade. Under this use the brick is not likely to be permeated with water. It can dry out easily.

Grade NW: This type of brick is intended for backup or for interior walls or if exposed to the outside where no frost action occurs.

Note: The word "permeated" as referred to in the ASTM (American Society for Testing and Materials) means the complete saturation of a brick with water where moisture is drawn through the brick by capillary action, wetting the brick from face to face. When a brick is completely immersed in water it will become saturated within a 24-hr period.

Facing brick. There are two grades of facing brick, based on (1) resistance to weather conditions and (2) factors based on appearance of the finished brick. Regarding the resistance to weather conditions, the same is true of these brick as for building brick grades SW and MW.

Regarding factors based on appearance there are the following types:

FBX: Where a high degree of mechanical perfection is desired, either on inside or outside walls

FBS: Where wide color ranges are desired; used for both interior and exterior masonry walls

FBA: a nonuniform-sized brick intended for architectural affect and variations

MORTAR MIXES REQUIRED TO LAY 1000 COMMON BRICKS

Kind of Mortar	Lime (lb)	Cement (bbl)	Sand (cu ft)
1:3 Lime mortar	180	—	18
2:1:9 Lime–cement mortar	115	0.50	18
1:3 Cement mortar, 10% lime	16	1.50	18
Brixment cement mortar	—	1.25	15
Masonry cement mortar	—	1.25	15
Kosmortar	—	1.25	15

Modular Brick

Modular brick has become the standard for most government agencies. The better architectural offices also specify the use of modular brick design. Modular brick are easier to take-off when a quantity survey is being prepared, are cheaper to lay, are coordinated with windows manufactured by members of the metal and wood institutes, and in their use eliminate cutting and fitting around openings, which represents a substantial saving on labor time.

On the basis of the 4-in. module taken both vertically and horizontally, certain sizes of brick have been manufactured to fit within the 4-in. increments.

In the following illustrations it can readily be seen that three $2\frac{2}{3}$-in. modular bricks with a $\frac{1}{2}$-in. mortar joint will fit into an 8-in. height, four 3-in. modular bricks will fit into a 12-in. height, and one 4-in. modular brick will fit into a 4-in. height.

A $2\frac{2}{3}$-in. modular brick actually measures $2\frac{1}{6}$ in. high, plus a $\frac{1}{2}$-in. mortar joint = $2\frac{2}{3}$-in.

A 3-in. modular brick actually measures $2\frac{1}{2}$ in. high, plus a $\frac{1}{2}$-in. mortar joint = 3 in.

A 4-in. modular brick actually measures $3\frac{1}{2}$-in. high, plus a $\frac{1}{2}$-in. mortar joint = 4 in.

In the first of the following illustrations the modular brick measures $2\frac{1}{6}$ in. $\times$ $3\frac{1}{2}$ in. $\times$ $7\frac{1}{2}$ in. If a $\frac{1}{2}$-in. mortar joint is used, three courses of brick lay up perfectly to every two horizontal grid lines, and one length of brick fits two vertical grid lines. The same holds true of the other sizes of brick shown.

NOMINAL MODULAR SIZES OF BRICK
NOMINAL SIZES INCLUDE THE THICKNESS OF THE STANDARD MORTAR JOINT

Thickness (in.)	Face dimensions in wall (in.)	
	Height	Length
4	2	12
4	$2\frac{2}{3}$	8
4	$2\frac{2}{3}$	12
4	4	8
4	4	12
4	$5\frac{1}{3}$	8
4	$5\frac{1}{3}$	12

Mortar

This is a mixture of cement, lime, and water. Use 1 part of cement to 3 parts of sand, to which is usually added some lime to make the mixture more workable.

A good brick mortar may consist of 1 part of cement, 1 part of lime, and 6

NUMBER OF MODULAR BRICK PER SQUARE FOOT

Size of brick including mortar joint	Face area of brick (sq in.)	Number of bricks (per sq ft)
$4 \times 2 \times 12$	24	4
$4 \times 2\frac{2}{3} \times 8$	21.28	6.76
$4 \times 2\frac{2}{3} \times 12$	31.92	4.51
$4 \times 4 \times 8$	32	4.5
$4 \times 4 \times 12$	48	3
$4 \times 5\frac{1}{3} \times 8$	42.64	3.4
$4 \times 5\frac{1}{3} \times 12$	63.96	2.26

parts of sand. Assuming the most commonly used brick joint of $\frac{1}{2}$ in. thickness, for both end and bed joints, about $16\frac{1}{2}$ cu ft of mortar is required for 1000 bricks.

Bricklayers will lay between 500 and 600 bricks per 8-hr day, and only about 300 or fewer face bricks per 8-hr day. About 600 sq ft of wall can be cleaned by the bricklayer in an 8-hr day.

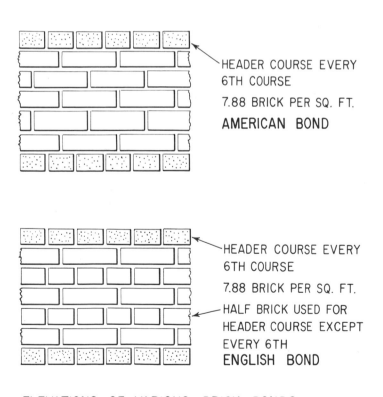

HEADER COURSE EVERY 6TH COURSE

7.88 BRICK PER SQ. FT.

AMERICAN BOND

HEADER COURSE EVERY 6TH COURSE

7.88 BRICK PER SQ. FT.

HALF BRICK USED FOR HEADER COURSE EXCEPT EVERY 6TH

ENGLISH BOND

ELEVATIONS OF VARIOUS BRICK BONDS

ALTERNATE FULL HEADERS
EVERY 6TH COURSE

7.15 BRICK PER SQ. FT.

HALF BRICK

FLEMISH BOND

ALTERNATE FULL HEADERS
EVERY 6TH COURSE
7.15 BRICK PER SQ. FT.
FLEMISH CROSS BOND

CONTINUOUS FULL HEADERS
EVERY 6TH COURSE
7.88 BRICK PER SQ. FT.
ENGLISH CROSS BOND

ELEVATIONS OF VARIOUS BRICK BONDS

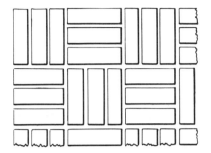

6.75 BRICK PER SQ. FT.
BASKET PATTERN

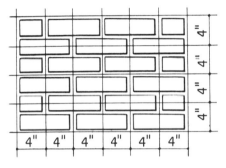

BRICK SIZE $2\frac{1}{6}'' \times 3\frac{1}{2}'' \times 7\frac{1}{2}''$

HEIGHT EACH COURSE $2\frac{2}{3}''$

6.75 BRICK PER SQ. FT.
(4" WALL)

$2\frac{2}{3}''$ **MODULAR BRICK**

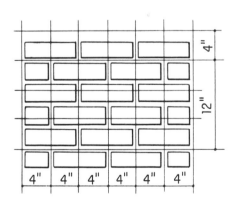

BRICK SIZE $2\frac{1}{2}'' \times 3\frac{1}{2}'' \times 7\frac{1}{2}''$

HEIGHT EACH COURSE 3"

6.00 BRICK PER SQ. FT.
(4" WALL)

3" **MODULAR BRICK**

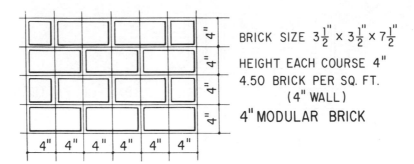

BRICK SIZE $3\frac{1}{2}'' \times 3\frac{1}{2}'' \times 7\frac{1}{2}''$

HEIGHT EACH COURSE 4"

4.50 BRICK PER SQ. FT.

(4" WALL)

4" MODULAR BRICK

Other Brick Sizes

Listed below are standard sizes most commonly manufactured by the brick industry in southern California. In addition, various manufacturers produce sizes other than these standards.

Name	Height (in.)	Width (in.)	Length (in.)
Standard building brick	$2\frac{1}{2}$	$3\frac{7}{8}$	$8\frac{1}{2}$
Oversize building brick	$3\frac{1}{4}$	$3\frac{1}{4}$	10
Modular building brick	$3\frac{3}{8}$	3	$11\frac{3}{8}$
Standard face brick	$2\frac{7}{16}$	$3\frac{1}{2}$	$7\frac{1}{2}$
Norman face brick	$2\frac{7}{16}$	$3\frac{1}{2}$	$11\frac{1}{2}$
Continental face brick	$3\frac{3}{8}$	3	$11\frac{1}{2}$
5-in. hollow brick	$3\frac{1}{2}$	$4\frac{1}{2}$	$11\frac{1}{2}$
Imperial, or Atlas brick	$5\frac{3}{8}$	3	$15\frac{3}{8}$
Padre brick	4	3	$15\frac{1}{2}$
	4	3	$11\frac{1}{2}$
	4	3	$7\frac{1}{2}$
"Four-16" brick	$3\frac{1}{2}$	3	$11\frac{1}{2}$
Royal brick block	$7\frac{5}{8}$	$5\frac{1}{2}$	$15\frac{1}{2}$
	$7\frac{5}{8}$	$7\frac{1}{2}$	$15\frac{1}{2}$
Paving brick	$2\frac{7}{16}$	$3\frac{1}{2}$	$7\frac{1}{2}$
Paving tile	$1\frac{1}{4}$	$3\frac{1}{2}$	$7\frac{1}{2}$
	$1\frac{1}{4}$	$3\frac{1}{2}$	$11\frac{1}{2}$
Bel Air paver	$2\frac{1}{4}$	$5\frac{1}{2}$	$11\frac{1}{2}$
Fire brick	$2\frac{1}{2}$	$4\frac{1}{2}$	9
Fire brick splits	$1\frac{1}{4}$	$4\frac{1}{2}$	9

REVIEW QUESTIONS

1. Explain how accurate brick estimating is accomplished.
2. How many Padre bricks with a ½-in. mortar joint are contained in 1 sq ft?
3. What percentage must be added to a common brick bond every sixth course?
4. Where is a grade SW brick used?
5. What is the average cost per M common brick, including material and installation?
6. What is the cost of a brick veneer wall 36 ft long and 4 ft high when there are 6.75 bricks in 1 sq ft?

7

CONCRETE BLOCK

INTRODUCTION

In this unit you will learn how to estimate the number of concrete blocks in a wall, together with the material and installation costs.

Concrete block is perhaps the most widely used masonry unit material in building construction today. Because of the ease in manufacture and its light weight, strength, and fireproofing qualities, it has supplanted many other types of masonry units. It is used for foundations, backup, fireproofing, load-bearing and non-load-bearing walls and partitions and, with special surface finishes, for exterior and interior areas where easy maintenance and stain resistance are required.

Concrete block units made in two-core design instead of three offer several advantages. One is the face-shell thickness at the center web, which increases the tensile strength of the unit and reduces the tendency of cracking due to drying, shrinkage, and temperature changes. Two-core block has three webs, reducing heat conduction, is about 4 lb lighter in weight, and allows more space in placing conduits or other utilities.

Grades of Concrete Block

Load-bearing concrete block units are classified in the American Society for Testing and Materials (ASTM) specifications as grade N and grade S, with grade N units suitable for general use, such as for exterior walls below and above grade, interior

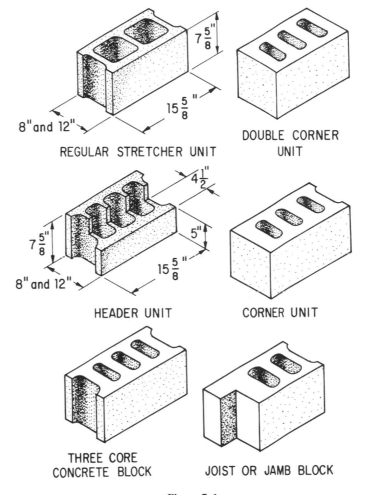

REGULAR STRETCHER UNIT

DOUBLE CORNER UNIT

HEADER UNIT

CORNER UNIT

THREE CORE CONCRETE BLOCK

JOIST OR JAMB BLOCK

Figure 7–1

walls, and backup. Grade S units are limited to use above grade in exterior walls with weather protective coatings and in walls not exposed to the weather.

Modular concrete blocks are the most frequently used blocks (Fig. 7-1). They measure $15\frac{5}{8}$ in. long by $7\frac{5}{8}$ in. high and are available in 8- and 12-in. thicknesses.

With a $\frac{3}{8}$-in. mortar joint, the length of a stretcher block from center to center of the mortar joint measures 16 in. Similarly, the height of the block, measured from center to center of the mortar joint, is 8 in.

A modular concrete block wall will therefore measure in length on a 16-in. module and in height on an 8-in. module (see Fig. 7-2). Window and door openings can be planed on a 16- or 8-in. module, eliminating the cutting and fitting of brick around openings.

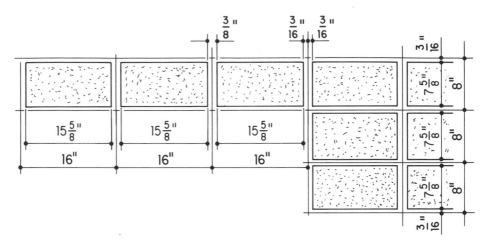

MODULAR CONCRETE BLOCK

Figure 7–2

THE TAKE-OFF

Concrete blocks are estimated by the number of units of each size required.   Areas of all openings should be deducted from the total wall surface.   Estimate the square feet of net wall, and then determine the number of blocks of various types required. This is done in the following manner.

Finding the Area of the Wall

From the outside perimeter of the wall, deduct four times the thickness of the wall and multiply by the height of the wall.   This is the area of the wall for masonry unit estimating.   The formula avoids doubling of corners.

For example, find the area of the wall in the plan and section of Fig. 7-3.

The outside perimeter of the wall is 20 ft + 20 ft + 10 ft + 10 ft = 60 lin ft.   Deduct four times the thickness of the wall (the wall is 1′0″).   Therefore,

$$60 \text{ ft} - 4 \text{ ft} = 56 \text{ lin ft}$$

Another method of finding the perimeter of the wall is by calculating the total stretch-out of the wall, which is 20 ft + 20 ft + 8 ft + 8 ft = 56 lin ft.   The perimeter with its four corner deductions, or the stretch-out, is then multiplied by the height, 6′0″ × 56′ = 336 sq ft decimal part of a foot in the following manner:

$$6'4'' = 6.33 \text{ ft}$$

$$\tfrac{1}{2}'' = \underline{0.04 \text{ ft}}$$

$$6'4\tfrac{1}{2}'' = 6.37 \text{ ft}$$

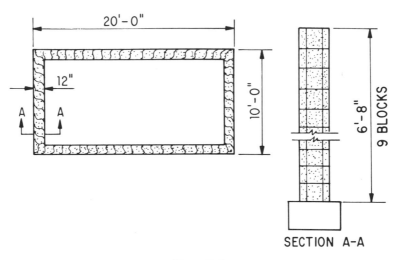

SECTION A-A

Figure 7–3

When the square feet of the net wall are known, refer to Fig. 7-4, which gives the number of blocks per 100 sq ft of wall. There are 110 blocks of 8 in. × 12 in. blocks in 100 sq ft of wall. The total square feet of wall is 336 sq ft.

$$336 \text{ sq ft} \div 100 = 3.36$$

$$110 \text{ blocks} = 100 \text{ sq ft}$$

$$336 \text{ sq ft} = 3.36 \times 110 = 370 \text{ blocks}$$

Finding the Number of Corner Blocks

The height of the wall (Fig. 7-3) is 6′0″ or 72 in. 72 in. ÷ 8 in. (height of one block) = 9 courses of blocks are required. Each corner has nine corner blocks.

$$\text{Four corners} = 4 \times 9 = 36 \text{ corner blocks}$$

$$370 \text{ (blocks)} - 36 \text{ (corner blocks)} = 334 \text{ stretcher blocks}$$

TABLE OF CONCRETE BLOCKS			
Size of Block	Wall Thickness	Number of Units per 100 sq. ft. of Wall	Mortar (cu. ft.)
8 x 8 x 16	8	110	3.25
8 x 8 x 12	8	146	3.50
8 x 12 x 16	12	110	3.25

Figure 7–4

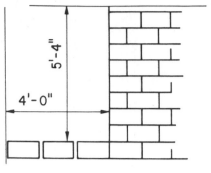

Figure 7–5   Figure 7–6

The amount of mortar required for blocks as determined trom Fig. 7-4 is

$$3.34 \times 3.25 \text{ cu ft} = 10.8 \text{ or } 11 \text{ cu ft (this includes 25\% mortar waste)}$$

Finding the Number of Jamb Blocks

Assume that two window openings on a plan having a masonry opening 4′0″ wide by 5′4″ high. How many jamb blocks will be required in Fig. 7-5?
The height of each block is 8 in. Therefore,

$$5'4'' = 64 \text{ in.} \div 8 = 8 \text{ blocks}$$

Each window has two rows of eight blocks. Two windows have $4 \times 8 = 32$ jamb blocks.

Finding the Number of Joist Blocks

The joist block is similar to the jamb block. It is used where wood joists are set into a concrete block wall, as shown in Fig. 7-6. To estimate the number of joist blocks for a wall, find the length of the wall on which the joists rest. Double this length if both ends of the joists rest on the block wall. Divide this total length by the length of one block to get the total number of joist blocks. The number of joist blocks are then subtracted from the total blocks of the wall. The remainder are the number of stretcher blocks.

    For example, a cellar is 26′0″ wide by 36′0″ long (outside dimensions). How many joist blocks are required? The block size is 8 in. × 12 in. × 16 in. The inside length of the cellar is 36 ft less 2 ft (twice the thickness of the wall), or 34 ft. The joists are spaced 16 in. on center to fit the pocket of each joist block. (The joist block is 16 in. long.) Then 34 ft × 12 = 408 in.

$$408 \div 16 = 25.5, \text{ or } 26, \text{ blocks}$$

$$26 \times 2 = 52 \text{ blocks}$$

Problem

For the cellar plan and section shown in Fig. 7-7, estimate the following: (1) the number of standard stretcher blocks, (2) the number of corner blocks, and (3) the number of window jamb blocks. *Note:* Stretcher blocks are 8 in. × 12 in. × 16 in.; window masonry openings are 1'4" high by 2'8" wide, and mortar joints are ½ in.

Solution Find the total area of the wall and subtract the area of window openings to get the net area of the wall. Let us find the area in tabular form:

Item	Unit	Length	Height	Square feet	Square feet out	Total square feet
North and south wall	2	74.67	5.33	795.88		
East and west wall	2	22.67	5.33	241.66		
Returns	2	4.00	5.33	42.64		
Windows (out)	4	2.67	1.33		14.20	
				1080.28		
				− 14.20		
				1066.08		1066

There are 110 blocks of 8 in. × 12 in. × 16 in. blocks in 100 sq ft of wall (see Fig. 7-4). The total square feet of wall is 1066.

$$1066 \div 100 \text{ sq ft} = 10.66 \text{ hundreds of square feet}$$

$$10.66 \times 110 = 1173 \text{ blocks}$$

Corner Blocks

The wall height is 5'4", or 64 in.

$$64 \text{ in.} \div 8 \text{ in. (height of block)} = 8$$

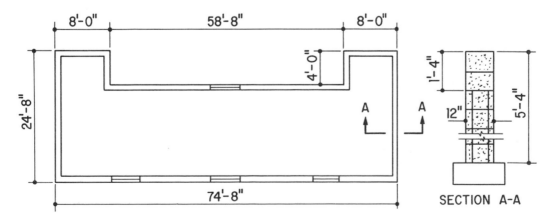

Figure 7–7

Therefore, eight courses of block are required. Each corner has eight corner blocks:

$$8 \times 8 = 64 \text{ corner blocks}$$

Jamb Blocks

The height of each block is 8 in. Therefore,

$$1'4'' = 16 \text{ in.} \div 8 = 2 \text{ blocks}$$

Each window has two rows of jamb blocks. Four windows have $8 \times 2 = 16$ jamb blocks. Then there are 64 corner blocks, 16 jamb blocks, or a total of 80 jamb and corner blocks, and 1173 total blocks less 80 = 1093 stretcher blocks.

MATERIAL AND INSTALLATION COSTS

Concrete block: 8 in. × 8 in. × 16 in., 75% solid, costs about $1.26 per square foot for the material and $1.90 for installation per square foot, or a total of $3.10 per square foot.

An 8 in. × 12 in. × 16 in. concrete block, 75% solid, costs about $3.65 per square foot for the material and the installation. A contractor's markup of 28% should be added for overhead and profit.

Lightweight concrete block: for partitions 8 in. × 16 in. × 4 in. thick costs $2.64 per square foot for material and installation. Assume an additional contractor's markup of 30% for overhead and profit.

Mortar for concrete block: 8 in. × 8 in. × 16 in. concrete block used as backup requires approximately 80 cu ft of mortar per M blocks. When the same block used as a partition, about 60 cu ft of mortar is required per M blocks.

8 in. × 12 in. × 16 in. concrete block used as partition requires approximately 90 cu ft, and used as backup about 110 cu ft of mortar is required.

SELF EXAMINATION

Solve the following problem and check your answers with those given in the Answer Box.

Estimate the number of concrete blocks used for the plan and section in Fig. 7-8.

1. Find the net square feet of wall.
2. Find the total number of blocks.
3. Find the number of corner blocks.

4. Find the number of window jamb blocks.
5. Find the number of joist blocks.
6. Find the number of stretcher blocks.

Note: Windows are $4'0'' \times 4'8''$.   All mortar joints are $\frac{3}{8}$ in.

Answer Box	
Square feet (net)	649
Total blocks	714
Corner blocks	64
Jamb blocks	56
Joist blocks	52
Stretcher blocks	542

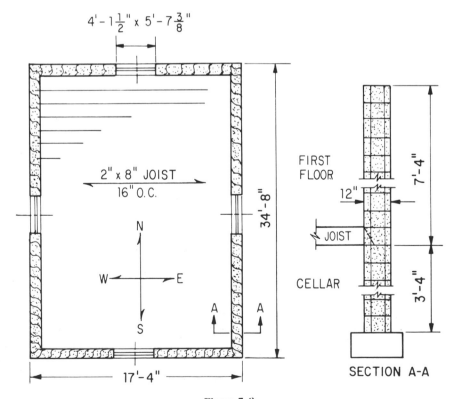

Figure 7–8

ASSIGNMENT—7

1. On $8\frac{1}{2}$ in. × 11 in. paper sketch the following plan and section on the upper half of the sheet. On the lower half, in tabular form, find the net square feet of wall for the plan and section in Fig. 7-9.

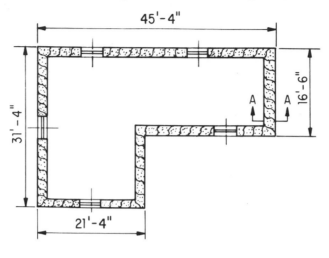

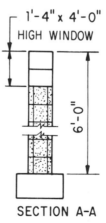

SECTION A-A

Figure 7–9

2. On another sheet of paper find (a) the number of corner blocks, (b) the number of jamb blocks, (c) the number of 8 in. × 12 in. × 16 in. stretcher blocks, and (d) the cubic feet of mortar ($\frac{3}{8}$-in. joint). *Note:* The window masonry openings are 1′4″ high by 4′0″ wide.

3. What is the material cost? The installation cost? The total cost, including subcontractor's overhead and profit?

SUPPLEMENTARY INFORMATION

CONCRETE BLOCKS AND MORTAR PER 100 SQ FT

Description, size of block (in.)	Wall thickness (in.)	Weight per unit (lb)	Number of units per 100 sq ft of wall area	Mortar (cu ft)	Weight (pounds per 100 sq ft of wall area)
8 × 8 × 16	8	50	110	3.25	5850
8 × 8 × 12	8	38	146	3.5	6000
8 × 12 × 16	12	85	110	3.25	9700
8 × 3 × 16	3	20	110	2.75	2600
9 × 3 × 18	3	26	87	2.5	2500
12 × 3 × 12	3	23	100	2.5	2550
8 × 3 × 12	3	15	146	3.5	2550
8 × 4 × 16	4	28	110	3.25	3450
9 × 4 × 18	4	35	87	3.25	3350
12 × 4 × 12	4	31	100	3.25	3450
8 × 4 × 12	4	21	146	4	3500
8 × 6 × 16	6	42	110	3.25	5000

REVIEW QUESTIONS

1. Give the dimensions of a regular modular stretcher concrete block.
2. How many units of 8 in. × 12 in. × 16 in. block are there in 100 sq ft of concrete block wall?
3. How would you find the number of jamb blocks required for a window?
4. An 8 in. × 16 in. block partition, 8 in. thick, costs how much per square foot installed?
5. What is the material and installation cost of an 8 in. × 12 in. × 16 in. block wall, 34′0″ long and 6′0″ high?   With a 30% overhead and profit?
6. How many cubic feet of mortar is required for the wall in Problem 5?

8

READING STEEL PLANS
AND SCHEDULES
AND ESTIMATING STEEL

INTRODUCTION

In this unit you will learn to read a typical partial structural foundation and first-floor plan, together with their footing and column schedules. This includes the take-off of steel angles used as lintels, steel beams and columns supporting structural slabs, bearing plates used for the proper seating of beams on concrete walls, and column base plates that receive the steel columns.

The Structural Foundation Plan

The partial structural foundation plan (Fig. 8-1) indicates concrete wall columns numbered 1, 2, 6, and 11. The dashed-line squares surrounding the columns represent the footings under the columns. Footings 7 and 12 support steel columns.

On the west wall between columns 1 and 2 are shown two beam pockets, which are formed in the concrete to receive the bearing ends of steel beams. Each pocket receives a steel bearing plate for the flush seating of the beam. The depth of the pocket is determined by the size of the beam. An 18 in.-high beam will have a deeper pocket or minus dimension than a 12 in.-high beam if the tops of all beams are to be kept at an even level. The width of the pocket must be several inches wider than the beam flange width.

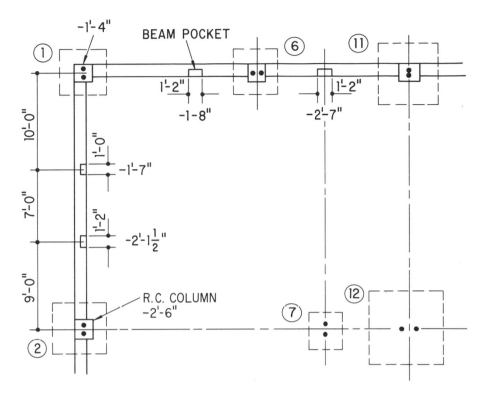

FOUNDATION PLAN

Figure 8–1

The Structural First-Floor Plan

In reading and taking-off the steel from the steel plan (Fig. 8-2), begin at the top of the plan and read the steel members consecutively from east to west. In this case, the first steel is a C5-6.7. This means a 5 in.-high channel weighing 6.7 lb per linear foot. The length of the channel is 2'6". The total weight of the channel is then 16.75 lb.

Further, the channel is located 4'0" from column 6. Since there are two channels, the total weight is $2 \times 16.7 = 33.5$ lb. The next steel member on the plan is a W6-9 followed by a W10-26, three W8-15, and so on.

After all east to west members are listed, the steel running north to south on the plan are then considered. The first two members are W12-16, followed by W12-50, two C7-9.8, W21-93, and W24-76. The weight of each member, per foot,

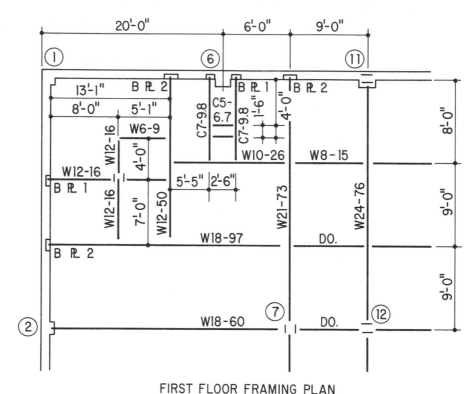

FIRST FLOOR FRAMING PLAN

Figure 8–2

is multiplied by its length to get the total pounds which is divided by 2000 to arrive at the total tonnage of steel.

Concrete Column Schedule

A typical reinforced concrete column schedule for columns 1, 2, 6, and 11 only is shown in Fig. 8-3. Column 1, for example, is 20 × 18 in. in size, has four No. 6 vertical reinforcings, and its lateral ties are No. 3 at 11 in. o.c.   The column elevation at the top is −1'4", and its minus dimension at the bottom is −12'8".

A typical detail of column 1 is shown in Fig. 8-4.   Note the vertical reinforcing bars 2 in. from the outer faces of the column and the lateral ties 11 in. o.c.   A ¼-in. plate is embedded in ½-in. cement grout and leveled before the grout hardens. Next, the base plate for the steel column is placed plumb and true to receive the column, which is welded to the base plate.   Two anchor bolts, ¾-in. in diameter, prevent the base plate from shifting on the setting plate.

REINFORCED CONCRETE COLUMN SCHEDULE					
Column Number	Column Size (in.)	Vertical Reinforcements	Lateral Ties	Top Elevation	Bottom Elevation
1	20 x 18	4- # 6	#3 11 in. o.c.	–1'4"	–12'8"
2	34 x 18	4- # 8	#3 12 in. o.c.	–2'6"	–12'8"
3					
4					
5					
6	18 x 17	4- # 6	#3 11 in. o.c.	–1'4"	–12'8"
10					
11	36 x 18	4- # 6	#3 12 in. o.c.	–3'0"	–12'8"
17					
18					
19					
20					

Figure 8–3

Figure 8-5 indicates a typical detail of how steel columns 7 and 12 are set on a reinforced concrete footing.

Steel Column and Footing Schedule

The schedule in Fig. 8-6 is that for column 1 only. To understand the information given on the schedule, a detailed drawing simulating the information on the schedule is presented.

For example, reading from the bottom up we see that column 1 has 16 No. 7 reinforcings running each way in the footing, for a total of 32 bars.

The elevation at the bottom of the footing is –14'0". This means 14 ft below elevation 0'0", which is the finished first-floor level. The footing size is 4'6" × 4'6" × 1'4". (See both the schedule and the detail.) The concrete column size was given on the reinforced concrete column schedule.

The column plate is 14" × 1¼" × 1'2". By multiplying these figures, the cubic inches of the plate are found. When multiplied by the weight per cubic inch, the total pounds of steel is found, which later is listed in tons.

The reinforced concrete column terminates at –1'4" from elevation 0'0". The bottom of the W8-35 column is welded to the base plate 1'1¼" from elevation 0'0".

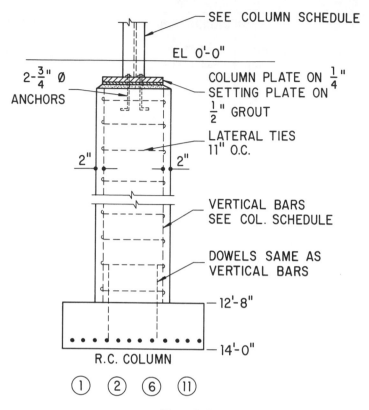

Figure 8–4

The W8-35 column is spliced to a W8-25 which terminates $2\frac{1}{2}$ in. from the top of the bulkhead slab, which is $+35'5''$.

In estimating the probable weight of structural steel for a job, the estimator should determine from the plans the total number of linear feet for each shape by

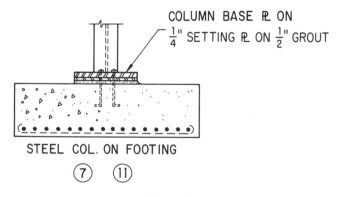

Figure 8–5

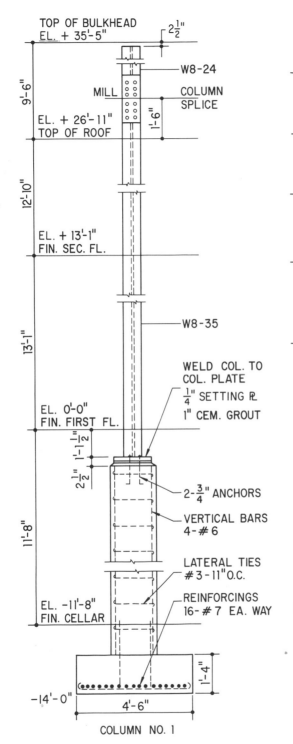

STEEL COL. AND FOOT'G SCHEDULE		
COLUMN NO.	1	2
TOP OF BULKHEAD SLAB + 35'-5" TOP OF ROOF SLAB EL. + 26'-11"	W8-24	
FIN. SECOND FL. EL. 13'-1"		
FIN. FIRST FL. EL. 0'-0"	W8-35	
FIN. CELLAR FL. EL. -11'-8"	R.C. COLUMN 1'-1$\frac{1}{2}$"	
COLUMN PLATE	14" x 1$\frac{1}{4}$" x 1'-2"	
CONC. FOOTING	4'-6" x 4'-6" x 1'-4"	
EL. AT BOTTOM	-14'-0"	
REINFORCINGS EACH WAY	16 - #7	
COLUMN NO.	1	2

COLUMN NO. 1

NOTE:
FIRST DIM'S OF COL. BASE ℙ, PARALLEL TO COL. FLANGES.

PARTIAL COL. SCHEDULE
SEE REINFORCED CONC. COL. SCHEDULE FOR COL. SIZE, VERTICAL BARS & LATERAL TIES.

Figure 8–6

113

size or weight.   Structural steel handbooks give the nominal weights of all sections.

However, variations in weights, amounting to $2\frac{1}{2}\%$ above or below the nominal weights, are permissible and may occur.   The purchaser is charged for the actual weight furnished, provided that the weight does not fall outside the permissible variation.

The weight of the steel used for connections should be estimated and priced separately if a detailed estimate is desirable.   In estimating the weight of a steel plate having an irregular shape, the weight of the rectangular plate from which the shape is cut should be used.   Steel weighs 490 lb/cu ft or 0.283 lb/cu in.

Preliminary Considerations

Before we begin with the actual take-off of the various steel items, it might be well to review some of the most common structural steel members used around the job. Figure 8-7 shows the common steel shapes.   These are steel angles, I-beams, wide-flange beams, H-columns, channels, and steel bearing plates.

A typical concrete floor slab with steel members in the slab is shown in Fig. 8-8.   The three steel members are (a) two 9-in. channels used around a well opening within the slab, (b) a wide-flange beam used as a floor beam, and (c) an I-beam used as a spandrel beam at an exterior wall.   The principal difference between the I-beam and the wide-flange beam is the width of the flange.   The flange of the wide-flange beam is wider than the flange of the I-beam.

The notation, C9-15 in Fig. 8-8 means that the height of the steel member is 9 in., the steel member is that of a channel, and the weight of the channel per

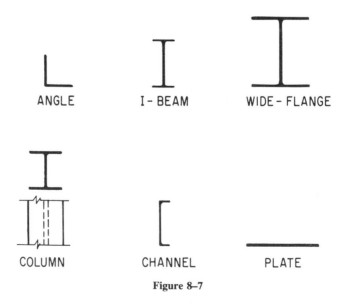

Figure 8–7

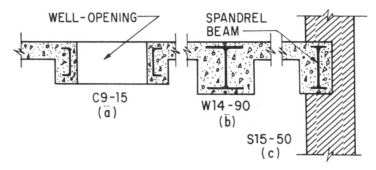

Figure 8–8

linear foot is 15 lb.   Similarly, the notation W14-90 means that the height of the member is 14 in. and that it is a wide-flange beam weighing 90 lb/lin ft.

Steel Lintels

These are steel angles used over windows and door openings in a brick wall; Fig. 8-9 will give you a good idea how steel angles are employed to support the bricks over the opening.   Imagine the part of the wall over the opening to be cut on line A-A.   The section you would see is as shown in B.   Other steel lintels in section are shown as in C, D, and E in Fig. 8-10.

THE TAKE-OFF

Steel lintels over windows and door openings are figured by multiplying the number of lintels over the openings by the length of the lintel by the weight of the lintel per linear foot to get the weight in pounds.

For example, find the total linear feet of steel angle lintels used over the windows and door openings on the plan in Fig. 8-11.   The 3'4" windows and the 2'0" windows require 3 at 4 in. × 3 in. × $\frac{5}{16}$ in. angles over each opening.   The two doors require 3 at 4 in. × 3 in. × $\frac{3}{8}$ in. angles over each opening.   Allow the minimum lintel bearing of 4 in. at each end of the opening.

$$4 \text{ (windows)} \times 3 \text{ (lintels)} \times 4'0'' \text{ (lintel length)} = 48 \text{ ft}$$

$$1 \text{ (window)} \times 3 \text{ (lintels)} \times 2.67 \text{ ft (lintel length)} = \underline{8.01 \text{ ft}}$$
$$56.01 \text{ ft}$$

Then

$$56.01 \text{ ft} \times 7.2 \text{ (weight of lintel in pounds per foot, Fig. 8-12)} = 403.27 \text{ lb}$$

$$2 \text{ (doors)} \times 3 \text{ (lintels)} \times 4.33 \text{ ft (length of lintel)} = 25.98 \text{ ft}$$

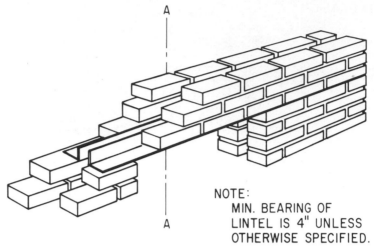

NOTE:
MIN. BEARING OF
LINTEL IS 4" UNLESS
OTHERWISE SPECIFIED.

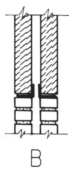

B

SECTION ON LINE A-A
OF LINTEL IN BRICK
CAVITY WALL

Figure 8–9

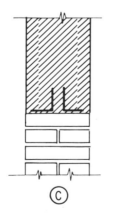

C

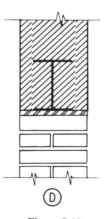

D

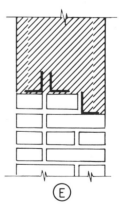

E

Figure 8-10

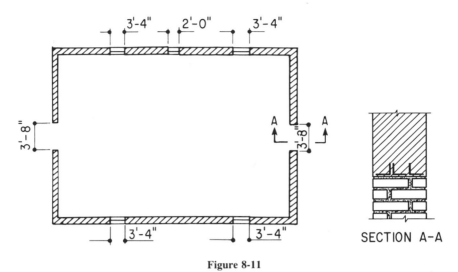

Figure 8-11

Then

25.98 ft $\times$ 8.5 (weight of lintel in pounds per foot, Fig. 8-6) = 254.83

$$\begin{array}{r} 403.27 \\ \underline{254.83} \\ 658.10 \end{array}$$ ÷ 2000 = 0.329 ton

Tabular Take-Off of Steel Lintels

Following is the tabular take-off of the lintels over the doors and windows of the preceding example:

STEEL LINTELS OVER OPENINGS

Item	Units	Length	Total length	Weight per foot (lb)	Total pounds	Tons
Windows: 4 in. $\times$ 3 in. $\times \frac{5}{16}$ in. lintels	12	4.0	48			
	3	2.67	$\underline{8.01}$			
			56.01	7.2	403.27	
Doors: 4 in $\times$ 6 in. $\frac{7}{16}$ in. lintels	6	4.33	25.98	8.5	$\underline{254.83}$	
					658.10	0.329

Steel Beams and Girder Take-Offs

Find the total length of beams of the same size and multiply the weight per foot to get the total pounds.   Divide the pounds by 2000 to get the total tons.

TABLE OF STEEL ANGLES AND THEIR WEIGHTS PER LINEAR FOOT					
Length of Legs (in.)	Thickness (in.)	Weight per foot (pound)	Length of Legs (in.)	Thickness (in.)	Weight per foot (pound)
8 x 8	$\frac{1}{2}$	26.4	$5 \times 3\frac{1}{2}$	$\frac{5}{16}$	8.7
	$\frac{3}{8}$	32.7		$\frac{3}{8}$	10.4
	$\frac{3}{4}$	38.9		$\frac{1}{2}$	13.6
	$\frac{7}{8}$	45.0		$\frac{3}{4}$	19.8
	1	51.0	4 x 3	$\frac{1}{4}$	5.8
	$1\frac{1}{4}$	56.9		$\frac{5}{16}$	7.2
6 x 6	$\frac{3}{8}$	14.9		$\frac{3}{8}$	8.5
	$\frac{1}{2}$	19.6		$\frac{1}{2}$	11.1
	$\frac{5}{8}$	24.2	$3\frac{1}{2} \times 3\frac{1}{2}$	$\frac{1}{4}$	5.8
	$\frac{3}{4}$	28.1		$\frac{5}{16}$	7.2
	$\frac{7}{8}$	33.1		$\frac{3}{8}$	8.5
	1	37.4	$2\frac{1}{2} \times 2\frac{1}{2}$	$\frac{3}{16}$	3.07
6 x 4	$\frac{3}{8}$	12.3		$\frac{1}{4}$	4.1
	$\frac{1}{2}$	16.2		$\frac{5}{16}$	5.0
	$\frac{3}{8}$	20.0		$\frac{5}{8}$	5.9
	$\frac{3}{4}$	23.6			

Figure 8-12

For example, the plan in Fig. 8-13 shows the following steel members:

1 at W18-86 girder 30′0″ long

6 at W8-21 beams 15′0″ long

4 at 56-12.5 beams 15′0″ long

All steel members have a 6-in. bearing surface. Take-off the total tonnage of steel required in tabular form.

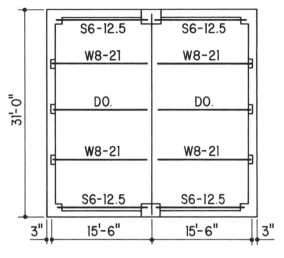

Figure 8-13

Item	Unit	Length	Total length	Weight per linear foot	Total weight (lb)	Tons
W18-86	1	30.0	30.0	85.0	2580	
W8-21	6	15.5	93.0	21.0	1953	
6 I-12.5	4	15.0	60.0	12.5	755	
					5288	2.64

Estimating the Weight of Bearing Plates and Column Base Plates

Steel bearing plates are used under the bearing ends of steel beams and girders on foundation walls. Column plates, anchored on top of reinforced concrete columns and footings, receive the steel columns. The steel columns are usually welded to the column plates.

To estimate the weight of such steel multiply the length of the plate in inches by the width of the plate in inches by the thickness of the plate in inches to find the cubic inches. Multiply the total cubic inches by 0.283 lb (the weight of the steel per cubic inch). Divide the total weight in pounds by 2000 (2000 lb equals 1 ton) to find the total tons of steel plate.

For example, find the total weight of steel plates required under the bearing ends of beams and on top of reinforced concrete columns and footings for the support of the steel columns shown in the plan in Fig. 8-14. The sizes of column plates for columns are shown on the column plate schedule.

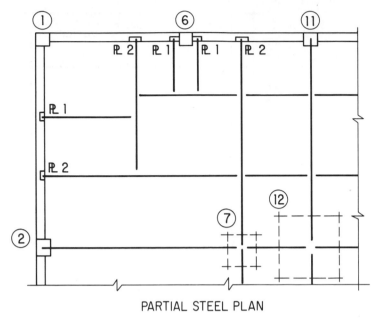

PARTIAL STEEL PLAN

NOTE: ℞ = BEARING PLATE

Figure 8-14

The sizes of the plates for beams are indicated on the following plate schedules:

COLUMN PLATE SCHEDULE

Column number	Column plate size
1	$14'' \times 1\frac{1}{4}'' \times 1'2''$
2	$32'' \times 3\frac{1}{2}'' \times 1'0''$
3	$24'' \times 2\frac{1}{2}'' \times 1'0''$
6	$14'' \times 1'' \times 1'0''$
7	$14'' \times 1\frac{1}{4}'' \times 1'0''$
11	$33'' \times 4'' \times 1'0''$
12	$29'' \times 3'' \times 2'5''$

BEARING PLATE SCHEDULE

Plate number	Bearing plate size
1	$8'' \times \frac{3}{4}'' \times 0'8''$
2	$8'' \times 1'' \times 1'0''$

To solve this problem, first find the cubic inches of steel in each column plate and bearing plate:

Column number	Column plate size	Cubic inches
1	$14'' \times 1\frac{1}{4}'' \times 1'2''$	245
2	$32'' \times 3\frac{1}{2}'' \times 1'0''$	1344
6	$14'' \times 1'' \times 1'0''$	168
7	$14'' \times 1\frac{1}{4}'' \times 1'0''$	210
11	$33'' \times 4'' \times 1'0''$	1584
12	$29'' \times 3'' \times 2'5''$	2523
		6074

Bearing plate number	Bearing plate size	Cubic inches	Units	Total cubic inches
1	$8'' \times \frac{3}{4}'' \times 0'8''$	48	2	96
2	$8'' \times 1'' \times 1'0''$	96	3	288
		144		384

Then

$$6794 \text{ cu in.} + 384 \text{ cu in.} = 7178 \text{ cu in.}$$

$$7178 \text{ cu in.} \times 0.283 \text{ lb/cu in.} = 2031.4 \text{ lb}$$

$$2031.4 \text{ lb} \div 2000 \text{ lb} = 1.015 \text{ ton}$$

The above can be taken-off in tabular form:

STEEL PLATES (0.283 LB/CU IN.)

Item	Unit	Width (in.)	Thickness (in.)	Length (in.)	Cubic inches	Pounds	Tons
Column plates							
1	1	14	1.25	14	245		
2	1	32	3.5	12	1344		
6	1	14	1	12	168		
7	1	14	1.25	12	210		
11	1	33	4	12	1584		
12	1	29	3	29	2523		
Bearing plates							
1	2	8	0.75	8	96		
2	3	8	1	12	288		
					6450		
					$\times 0.283$	1825.35	0.91

MATERIAL AND INSTALLATION COSTS

Structural steel worker: $23.75 per hour

Steel in-place: General average per ton of steel in place for A36 high-strength steel is $916; including subcontractor's overhead and profit, $1371.

a. Material cost per ton:

Base price at mill	$362
Delivery to shop	60
Drafting (shop drawings)	38
Shop fabrication	242
Shop coat paint	25
Trucking to job	18
	$745

b. Installation cost per ton:

Unload	19
Erect and plumb	88
Field bolts	31
Crane erection (include operator)	33
	$171
Total per ton in-place	$916

Steel lintels—plain angles:

Under 500 lb: $0.71 per pound

500 to 1000 lb: $0.62 per pound

Steel lintels—plain angles with plates:

Under 500 lb: $0.83 per pound

500 to 1000 lb: $1.02 per pound

Steel lally column: concrete filled, 4-in. diameter, $11.05 per linear foot.

Steel beams: C10-20 cost $14.88 per linear foot installed; HP10-42 costs $21.45 per linear foot installed; W33-201 costs $40.45 per linear foot installed.

SELF EXAMINATION

Take-off the steel lintels required over the exterior and interior openings on the plan in Fig. 8-15.   Lintels shall be two steel angles 4 in. × 3 in. × $\frac{3}{8}$ in. over each opening.   Bearing of lintels shall be 4 in.   Prepare the take-off in tabular form.   Check your answer with that given in the Answer Box.   If you have difficulty, review this unit.

Answer Box
Windows = 66.64 ft
Doors = <u>14.68</u> ft
Total length = 81.32 ft
Total pounds = 691.22
Tons = 0.345

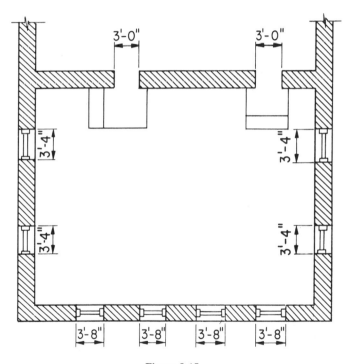

Figure 8-15

ASSIGNMENT—8

Take-off the steel beams and bearing plates shown on the plan in Fig. 8-16. Take-off the beams in tabular form and find the total tonnage of steel. Take-off the bearing plates in tabular form and find the total tonnage.

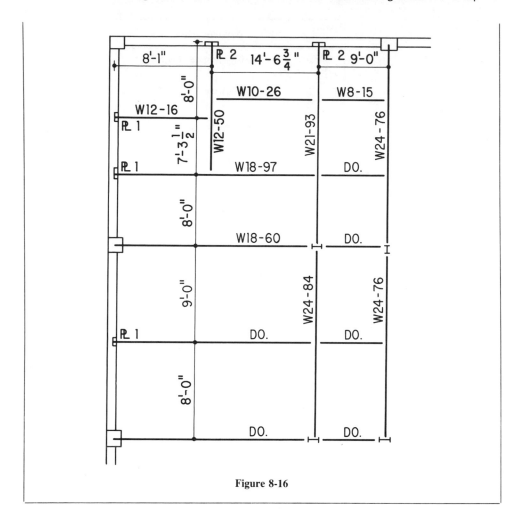

Figure 8-16

REVIEW QUESTIONS

1. What is the height of a reinforced concrete column with a top minus dimension of 1′6″ and a bottom minus dimension of 16′6″?
2. Briefly explain the function of the setting plate.
3. How are column plates estimated?
4. What is the general average cost per ton of steel for A36 high-strength steel in place?
5. Briefly explain how steel angle lintels are estimated.

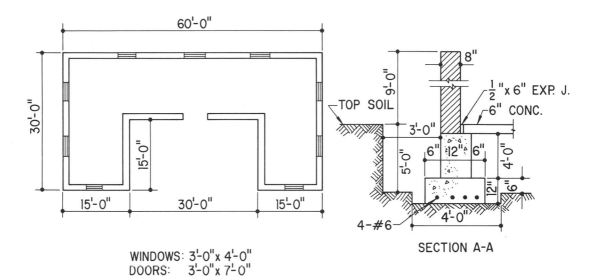

Figure 8-17

REVIEW PROBLEM OR MIDTERM EXAM

Take-off the following (Fig. 8-17):
1. General excavation.
2. Excavation for footing.
3. Concrete footing.
4. Concrete foundation wall.
5. Reinforcings in footing.
6. Forms for footing:
   (a) Contact forms 2 in. × 12 in.
   (b) Stakes 2 in. × 2 in. × 4 in. o.c.   2′0″ long.
   (c) Spreader ties 1 in. × 3 in. 4′0″ o.c.
   (d) Nails.
7. 6 in. concrete slab.
8. $\frac{1}{2}$ in. × 6 in. expansion joint.
9. Face and common: brick size $2\frac{1}{4}$ in. × $3\frac{1}{4}$ in. × 8 in., mortar joint $\frac{1}{2}$ in.

9

LUMBER

INTRODUCTION

In this unit you will learn how to compute the quantity of rough lumber in board feet for floor joists, ceiling beams, rafters, studs, and other framing members.

Lumber is the general term for the material used in carpentry work. Large-cross-section pieces are called *timbers*. Grades of lumber have been established by the trade associations. Fir, pine, and spruce are used for framing lumber and sheathing; while oak, maple, and white pine are ordinarily used for flooring and finish work. Boards are usually 1 in. thick and 2 to 10 in. wide. Boards less than 1 in. thick are figured as 1 in. for estimating purposes. Tongue-and-groove sheathing boards are considered 1 in. × 4 in., 1 in. × 6 in., and 1 in. × 8 in. but measure less than these dimensions.

Lumber that is used for general building purposes is called *yard lumber*. It is less than 5 in. in thickness. Yard lumber is classified into two main divisions on the basis of quality:

1. *Select lumber:* that which is generally clear, contains no defects or only a limited number with respect to both size and type, and is smoothly finished and suitable for use as a whole for finishing purposes or where large clear pieces are required. It is further classified by grades A, B, C, and D. Grades A and B are suitable for natural finish; grades C and D have defects and

blemishes of somewhat greater extent than A and B but which can be covered by paint.

2. *Common lumber:* this is classified by the American Lumber Standards Association as lumber having defects and blemishes not suitable for finishing purposes, but quite satisfactory for general utility and construction purposes. Lumber under this heading is further classified into grades 1, 2, 3, 4, and 5 common.

In grades 1 and 2 common, all defects and blemishes must be sound. Such lumber is used for purposes in which surface covering or strength is required. Grades 3, 4, and 5 common contains very coarse defects that will cause waste within a piece. Such lumber is used where strength is not important.

In rough carpentry, lumber is figured in board feet. We saw in Unit 4 that 1 B.F. = 144 cu in. of wood. By writing the standard dimension of lumber over 12 in the form of a fraction, the number of board feet of lumber is determined.

For example, a 2 in. × 12 in. floor joist 20 ft long, equals

$$\frac{2 \text{ in.} \times \cancel{12} \text{ in.} \times 20 \text{ ft}}{\cancel{12}} = 40 \text{ B.F.}$$

Where several pieces are required the number of board feet in a piece is multiplied by the number of pieces involved, such as

$$\frac{10 \text{ pc.} \times 2 \text{ in.} \times 12 \text{ in.} \times 20 \text{ ft}}{12} = 400 \text{ B.F.}$$

Finding the Cost per Board Foot

When it is necessary to find the cost per board foot and the cost per 1000 B.F. is known, divide the cost by 1000. When the cost per 100 B.F. is known, divide by 100 to get the cost per board foot.

For example, find the cost of 60 pieces of 2 in. × 8 in. × 20 ft at 1000 B.F.

$$\frac{2 \text{ in.} \times 8 \text{ in.} \times 20 \text{ ft} \times \overset{5}{\cancel{60}}}{\cancel{12}} = 1600 \text{ B.F.}$$

$$\$480.00 \div 1000 = \$0.48, \text{ cost per board foot}$$

$$1600 \text{ B.F.} \times 0.48 = \$768.00, \text{ cost of } 1600 \text{ B.F.}$$

Problem

Find the cost of the following lumber when 1000 B.F. costs $480.

20 pc. of 2 in. × 8 in. × 10 ft long

60 pc. of 2 in. × 8 in. × 8 ft long

30 pc. of 2 in. × 8 in. × 12 ft long

Solution

$$\frac{20 \times 2 \text{ in.} \times \overset{2}{\cancel{8}} \text{in.} \times 10 \text{ ft}}{\underset{3}{\cancel{12}}} = \frac{800}{3} = 266\tfrac{2}{3}, \text{ or } 267, \text{ B.F.}$$

$$\frac{\overset{5}{\cancel{60}} \times 2 \text{ in.} \times 8 \text{ in.} \times 8 \text{ ft}}{\cancel{12}} = 640 \text{ B.F.}$$

$$\frac{30 \times 2 \text{ in.} \times 8 \text{ in.} \times \cancel{12}\text{ ft}}{\cancel{12}} = 480 \text{ B.F.}$$

267 B.F. + 640 B.F. + 480 B.F. = 1387 total number of board feet

1387 B.F. × 0.48 = $665.76 total cost

Practice Problem

Find the cost of the following when 1000 B.F.   costs $492.

16 pc. of 2 in. × 6 in. × 8 ft long (128 B.F.)

36 pc. of 2 in. × 8 in. × 18 ft long (864 B.F.)

55 pc. of 2 in × 4 in. × 10 ft long (366⅔ B.F.)

Total cost = 668.46 (*ans.*)

Figuring the Number of Pieces within a Given Number of Board Feet

When a certain job calls for a given number of board feet of lumber and the price has been determined, it will be necessary in many instances to find the number of pieces of a certain size required.

For example, find the number of pieces of 2 in. × 10 in. × 18 ft long that can be delivered for an order of 2700 B.F.

$$\frac{\cancel{2} \text{ in.} \times 10 \text{ in.} \times \overset{3}{\cancel{18}} \text{ ft}}{\underset{\cancel{6}}{\cancel{12}}} = 30 \text{ B.F.}$$

2700 B.F. ÷ 30 B.F. = 90 pc. of 2 in. × 10. in × 18 ft long

Practice Problem

Find the number of 3 in. × 6 in. × 10 ft pieces that can be gotten out of 3600 B.F. (*Ans.*: 240 pc.)

Finding the Linear Feet of Lumber for a Given Number of Board Feet

Smaller-sized lumber is usually figured in linear feet. If 165 B.F. of 1 in. × 3 in. cross bridging is called for and it is necessary to find the total linear feet, the procedure is as follows:

$$\frac{1 \text{ in.} \times \cancel{3} \text{ in.}}{\underset{4}{\cancel{12}}} = \tfrac{1}{4} \text{ or } 0.25 \text{ B.F. in each linear foot of material}$$

$$\frac{165 \text{ B.F.}}{0.25 \text{ B.F.}} = 660 \text{ lin ft of material required}$$

Practice Problem

Find the linear feet of 1 in. × 2 in. furring that can be gotten from 3500 B.F. How many 8-ft. lengths? (*Ans.*: 21,124 lin ft or 2641 pc of 8-ft, length).

THE TAKE-OFF

Estimating the Number of Wood Joists

To find the number of floor joists over a certain area, divide the distance to be covered by the joists by the joist spacing and add one joist.

Problem

An area of 17'0" × 35'0" (Fig. 9-1) is to be covered by floor joists that are spaced 16 in. on center (o.c.). How many joists are required?

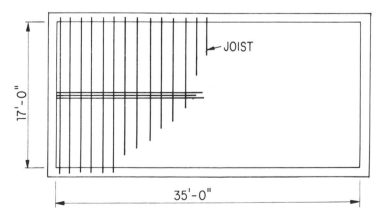

Figure 9-1

Solution

16-in. joist spacing = 1.33 ft

35 ft ÷ 1.33 ft = 26.3 or 27 + 1 = 28 joists

Whenever a fraction of a space results, it is customary to allow the full space.   The extra joist allowed is the one at the end of the span.   We can say that there is always one more joist than there are joist spaces.

Practice Problem

Floor joists, 2 in. × 8 in., cover a span of 50 ft.   The joists are spaced 16 in. o.c.   Find the number of joists.   (*Ans.*: 39.)   Find the number of board feet when joist lengths are 12 ft.   (*Ans.*: 624 B.F.)

Figuring Common Rafters

To find the number of common rafters, divide the rafter spacing by the length of the distance to be covered by the rafters and add one rafter.   Double the number thus found for two slopes of roof.

Problem

An area of 12 ft × 40 ft of two slopes is to be covered by roof rafters spaced 16 in. o.c. (Fig. 9-2).   How many 2 in. × 8 in. × 12 ft rafters are required?   What is the cost of the lumber when the price of 2 in. × 8 in. is $390 per 1000 B.F.?

Solution

16 in. rafter spacing = 1.33 ft rafter spacing

40 ft ÷ 1.33 ft = 30 spaces + 1 = 31 rafters

31 × 2 = 62 rafters (both sides of roof)

The total length of all rafters is 62 × 12 ft = 744 lin ft.

$$\frac{2 \times 8 \times \overset{62}{\cancel{744}}}{\cancel{12}} = 992 \text{ B.F.}$$

390 ÷ 1000 = $0.39/B.F.

992 B.F. × 0.39 = $386.88, price of the rafters

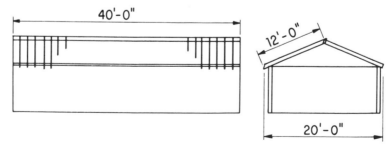

Figure 9-2

Estimating the Number of Wall Studs

The quickest and perhaps the best way to estimate the number of studs in the exterior and interior walls is to add the lengths of all walls running east and west and the walls running north and south, including door and window openings. Also include cased openings, which are indicated by dashed lines on plans. Since additional studs are required for constructing corner posts and framing around openings for doors, windows, and so on, with stud spacing of 16 in. on center, a common practice is to consider 1 stud/ft for the total linear feet of walls. Therefore, for estimating purposes, the total number of linear feet of outside and inside walls gives the number of studs required.

For example, find the number of 2 in. × 4 in. studs required for the outside walls and inside partitions of the four-room bungalow in Fig. 9-3. Assume a stud spacing of 16 in. on center and a stud height of 8′0″. The fireplace facade in Fig. 9-3 extends to the ceiling. (No allowance for studs is to be made for this opening.) In taking-off the linear feet of studding on the plan in Fig. 9-3 the circled numbers 1-8 point to walls that run west to east, while the circled letters A-F point to the walls that run north to south. Their lengths are computed as follows:

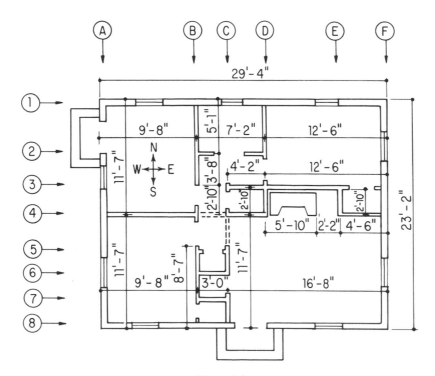

Figure 9-3

1.	29'4"	A	23'2"
2.	7'2"	B	23'2"
3.	16'8"	C	14'5"
4.	23'6"	D	11'7"
5.	3'0"	E	2'10"
6.	3'0"	F	23'2"
7.	3'0"		98'4"
8.	29'4"		
	115'0"		

115'0" + 98'4" = 213'4", or 214, ft, the total length of walls, which is also the number of studs required. If the outside walls require a different length of studding from the inside walls, the number of studs for the outside walls and inside partitions must be found separately.

Estimating the lumber required for the sole plate (shoe) and top plate (2 × 4's over studding). The total linear feet of outside walls and partition walls is the total linear feet of top plate and sole plate. This length is multiplied by 3 when two top plates are used and one sole plate.

For example, in the previous problem we found a total of 214 lin ft. Assuming two top plates and one sole plate, the total length of 2 × 4's is

$$214 \text{ ft} \times 3 = 642 \text{ lin ft}$$

If 12-ft lengths are desired,

$$642 \text{ ft} \div 12 \text{ ft} = 53.5, \text{ or } 54 \text{ pc. of}$$

$$2 \text{ in.} \times 4 \text{ in.} \times 12 \text{ ft lengths are required}$$

Insulating Sheathing

There are many brands of insulating sheathing boards on the market today that combine strength and resistance to moisture and air infiltration and have sound-deadening qualities. The boards, or "panels," are manufactured in sheets of various thicknesses and sizes. Common sizes for exterior wall sheathing panels are 4 ft. × 8 ft. and 4 ft. × 9 ft. with thicknesses of $\frac{1}{2}$ in. and $\frac{25}{32}$ in. Other sizes of sheets are available.

To estimate the quantity of this material, find the total square feet of surface covered, deducting window and door openings. The total net area is divided by the square feet of the panel size selected to get the number of panels for the job. No waste need be considered.

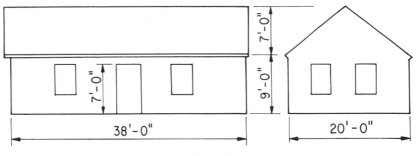

Figure 9-4

Plywood Sheathing Panels

Plywood sheathing panels are ordinarily 4 ft. × 8 ft. with thicknesses of $\frac{5}{16}$, $\frac{3}{8}$, $\frac{1}{2}$, $\frac{5}{8}$, and $\frac{3}{4}$ in. Figure the total net area to be covered and divide by 32 sq ft (the area of one 4 ft × 8 ft panel) to arrive at the number of panels required.

For example, find the total exterior wall sheathing surface of the house in Fig. 9-4, using plywood panels 4 ft × 8 ft. How many panels are required? *Note*: The window sizes are 2′6″ × 4′0″; the door size is 3′0″ × 7′0″. There are two windows on each end elevation and three windows on the rear elevation. Therefore,

$$38 \text{ ft} + 20 \text{ ft} + 38 \text{ ft} + 20 \text{ ft} = 116 \text{ ft outside perimeter}$$

$$116 \text{ ft} \times 9 \text{ ft (height of wall)} = 1044 \text{ sq ft of wall surface}$$

$$\text{Window area} = 2.5 \text{ ft} \times 4 \text{ ft} = 10 \text{ sq ft} \times 9 \text{ (windows)} = 90 \text{ sq ft}$$

$$\text{Door area} = 3 \text{ ft} \times 7 \text{ ft} = 21 \text{ sq ft}$$

$$\text{Windows and door} = 90 \text{ ft} + 21 \text{ ft} = 111 \text{ sq ft}$$

$$1044 \text{ sq ft} - 111 \text{ sq ft} = 933 \text{ sq ft (net)}$$

$$\text{Area of gables} = (7 \text{ ft} \times 20 \text{ ft} \times \tfrac{1}{2}) \times 2$$

$$= 140 \text{ sq ft. both ends}$$

By taking a rectangle, as in Fig. 9-5, both ends of the gable area are included. The two shaded triangles A and B are of the same area as the gable at the other end of the house. Then

$$933 \text{ sq ft} + 140 \text{ sq ft} = 1073 \text{ sq ft (net)}$$

$$4 \text{ ft} \times 8 \text{ ft panel} = 32 \text{ sq ft}$$

$$1073 \text{ sq ft} \div 32 \text{ sq ft} = 33.53, \text{ or } 34, \text{ panels}$$

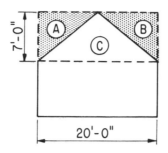

Figure 9-5

Problem

Estimate the quantities of lumber in Fig. 9-6, and determine the cost of the material when the cost per 1000 (M) B.F. is known.

Sill plate = 2 pc. of 2″ × 6″ at $380 per M B.F.

Built-up girder = 3 pc. of 2″ × 8″ at $390 per M B.F.

Floor joists = 2 in. × 8 in. at $390 per M B.F.

Headers = 2 in. × 8 in. at $390 per M B.F.

Solution

TABULAR TAKE-OFF

Item	Units	Length (ft)	Total length	Board feet	Cost per board foot	Total cost
2 in. × 6 in. sill	1	124	124	124	$0.38	$ 47.12
2 in. × 8 in girder	3	20	60	80	0.38	31.20
2 in. × 8 in. joists	31	12	372			
	16	12	192			
			564	752	0.38	293.28
2 in. × 8 in. header	2	40	80	106.66	0.38	41.59
						413.19

MATERIAL AND INSTALLATION COSTS

The costs for lumber are:

Joists—2 in. × 6 in. fir:

$380 per M fbm for material
 225 per M fbm for installation
$605 per M fbm

With subcontractor's overhead and profit, $726 per M fbm.

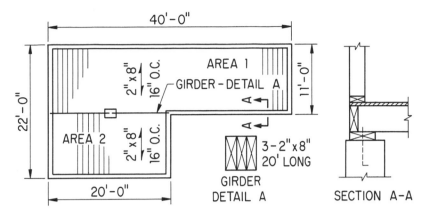

Figure 9-6

Joists—2 in. × 8 in. fir:

> $390 per M fbm for material
> <u>190</u> per M fbm for installation
> $580 per M fbm

With subcontractor's overhead and profit, $696 per M fbm.

Rafters: 4-to-12 pitch, 2 in. × 6 in.

> $330 per M fbm for material
> <u>280</u> per M fbm for installation
> $610 per M fbm

With subcontractor's overhead and profit, $732 per M fbm.

Studs: 8-ft-high wall, 2 in. × 4 in.

> $340 per M fbm for material
> <u>440</u> per M fbm for installation
> $780 per M fbm

With subcontractor's overhead and profit, $1053 per M fbm.

Roof sheathing: The cost for plywood, $\frac{1}{2}$ in. thick, is

> $0.34 per square foot for material
> <u>0.17</u> per square foot installation
> $0.51 per square foot

With subcontractor's overhead and profit, $0.61 per square foot.

Wall sheathing: The cost for plywood, $\frac{5}{8}$ in. thick, is

$0.44 per square foot for material
 0.23 per square foot for installation
$0.67 per square foot.

With subcontractor's overhead and profit, $0.79 per square foot.

Sill plate:

$520 per M fbm for material
 305 per M fbm for installation
$825 per M fbm

With subcontractor's overhead and profit, $974 per M fbm.

Mud sills: construction grade, 2 in. × 4 in.

$ 900 per M fbm for material
  395 per M fbm for installation
$1295 per M fbm

With subcontractor's overhead and profit, $1525 per M fbm.

Mud sills: construction grade, 2 in. × 6 in.

$ 900 per M fbm for material
  305 per M fbm for installation
$1205 per M fbm

With subcontractor's overhead and profit, $1400 per M fbm.

SELF EXAMINATION

Answer the following questions and check your answers with those given in the Answer Box.

1. A 2 in. × 12 in. floor joist 12 ft long contains how many board feet?
2. Find the cost of 48 pc. of 2 in. × 8 in. by 18 ft at $380 per 1000 B.F.
3. Find the cost of the following lumber when 1000 B.F. costs $380.

   30 pc. of 2 in. × 6 in. × 12 ft long

   45 pc. of 2 in. × 8 in. × 10 ft long

   60 pc. of 2 in. × 8 in. × 12 ft long

4. Find the number of pieces of 2 in. × 8 in. × 16 ft long that can be delivered for an order of 3200 B.F.
5. How many linear feet of 1 ft × 3 ft bridging are contained in 210 B.F.?
6. Assume the length of the plan in Fig. 9-1 to be 58 ft. How many joists would be required when the spacing is 16 in. o.c.?

Answer Box	
Problem	
1	24 B.F.
2	$437.76
3	$729.60
4	150 pc.
5	840 lin ft
6	45

ASSIGNMENT—9

1. Estimate the lumber required for the plan given in Fig. 9-7:

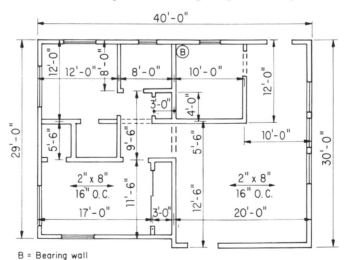

B = Bearing wall

Figure 9-7

   (a) Board feet of single sill (2 in. × 6 in.).
   (b) How many floor joists are required, and of what lengths?
   (c) Board feet of floor joist.
   (d) Linear feet of sole and two top plates.
   (e) Studding for exterior walls and partitions. Studs 8 ft long. Find the number of pieces.

2. Find the cost of studding when 1000 B.F. costs $780. Add 20% for overhead and profit.

REVIEW QUESTIONS

1. A 2 in. × 12 in. floor joist 24 ft long has how many board feet?

2. How many pieces of 2 in. × 8 in. × 16 ft can be delivered from an order of 2800 B.F.?

3. All outside and inside stud wall lengths add up to 413 ft.   How many studs are required? How many B.F. if the studs are 8 ft high?   What is the total length of 2 in. × 4 in. top and sole plate?

4. Explain how floor joists are estimated.

5. What is the cost of material and installation for the stud work in Problem 3?

SUPPLEMENTARY INFORMATION

RAPID LUMBER COMPUTATIONS IN FOOT BOARD MEASURE FOR GIVEN LENGTHS

Size (in.)	Length (ft)							
	8	10	13	14	16	18	20	22
1 × 2	$1\frac{1}{3}$	$1\frac{2}{3}$	2	$2\frac{1}{3}$	$2\frac{2}{3}$	3	$3\frac{1}{3}$	$3\frac{2}{3}$
1 × 3	2	$2\frac{1}{2}$	3	$3\frac{1}{2}$	4	$4\frac{1}{2}$	5	$5\frac{1}{2}$
1 × 4	$2\frac{2}{3}$	$3\frac{1}{3}$	4	$4\frac{2}{3}$	$5\frac{1}{3}$	6	$6\frac{2}{3}$	$7\frac{1}{3}$
1 × 5	$3\frac{1}{3}$	$4\frac{1}{6}$	5	$5\frac{5}{8}$	$6\frac{2}{3}$	$7\frac{1}{2}$	$8\frac{1}{3}$	$9\frac{1}{6}$
1 × 6	4	5	6	7	8	9	10	11
1 × 8	$5\frac{1}{3}$	$6\frac{2}{3}$	8	$9\frac{1}{3}$	$10\frac{2}{3}$	12	$13\frac{1}{3}$	$14\frac{2}{3}$
1 × 10	$6\frac{2}{3}$	$8\frac{1}{3}$	10	$11\frac{2}{3}$	$13\frac{1}{3}$	15	$16\frac{2}{3}$	$18\frac{1}{3}$
1 × 12	8	10	12	14	16	18	20	22
2 × 4	$5\frac{1}{3}$	$6\frac{2}{3}$	8	$9\frac{1}{3}$	$10\frac{2}{3}$	12	$13\frac{1}{3}$	$14\frac{2}{3}$
2 × 6	8	10	12	14	16	18	20	22
2 × 8	$10\frac{2}{3}$	$13\frac{1}{3}$	16	$18\frac{2}{3}$	$21\frac{1}{3}$	24	$26\frac{2}{3}$	$29\frac{1}{3}$
2 × 10	$13\frac{1}{3}$	$16\frac{2}{3}$	20	$23\frac{1}{3}$	$26\frac{2}{3}$	30	$33\frac{1}{3}$	$36\frac{2}{3}$
2 × 12	16	20	24	28	32	36	40	44
4 × 4	$10\frac{2}{3}$	$13\frac{1}{3}$	16	$18\frac{2}{3}$	$21\frac{1}{3}$	24	$26\frac{2}{3}$	$29\frac{1}{3}$
4 × 6	16	20	24	28	32	26	40	44
4 × 8	$21\frac{1}{3}$	$26\frac{2}{3}$	32	$37\frac{1}{3}$	$42\frac{2}{3}$	48	$53\frac{1}{3}$	$58\frac{2}{3}$
4 × 10	$26\frac{2}{3}$	$33\frac{1}{3}$	40	$46\frac{2}{3}$	$53\frac{1}{3}$	60	$66\frac{2}{3}$	$73\frac{1}{3}$
6 × 6	34	30	36	42	48	54	60	66
6 × 8	32	40	48	56	64	72	80	88
6 × 10	40	50	60	70	80	90	100	110
8 × 8	$42\frac{2}{3}$	$53\frac{1}{3}$	64	$74\frac{2}{3}$	$85\frac{1}{3}$	96	$106\frac{2}{3}$	$117\frac{1}{3}$
8 × 10	$53\frac{1}{3}$	$66\frac{2}{3}$	80	$93\frac{1}{3}$	$106\frac{2}{3}$	120	$133\frac{1}{3}$	$146\frac{2}{3}$
8 × 12	64	80	96	112	128	144	160	176

RAPID LUMBER COMPUTATIONS IN FOOT BOARD MEASURE

1 × 3	Divide linear feet by 4
1 × 4	Divide linear feet by 3
1 × 6	Divide linear feet by 2
1 × 8	Multiply linear feet by 2, and divide by 3
1 × 10	Multiply linear feet by 10, and divide by 12
1 × 12	Linear feet and foot board measure the same
2 × 3	Divide linear feet by 2
2 × 4	Multiply linear feet by 2, and divide by 3
2 × 8	Add to linear feet one-third the amount
2 × 10	Multiply linear feet by 10, and divide by 6
2 × 12	Multiply linear feet by 2
3 × 3	Multiply linear feet by 3, and divide by 4
3 × 4	Linear feet and foot board measure the same
3 × 6	Add to linear feet one-half the amount
3 × 8	Multiply linear feet by 2
3 × 10	Multiply linear feet by 10, and divide by 4
3 × 12	Multiply linear feet by 3
4 × 4	Add to linear feet one-third the amount
4 × 6	Multiply linear feet by 2
4 × 8	Multiply linear feet by 3, and subtract $\frac{1}{3}$ lin. ft. from amount
4 × 10	Multiply linear feet by 10, and divide by 3
4 × 12	Multiply linear feet by 4
8 × 8	Multiply linear feet by $5\frac{1}{3}$
10 × 10	Multiply linear feet by 100, and divide by 12
12 × 12	Multiply linear feet by 12
14 × 14	Multiply linear feet by $16\frac{1}{3}$

BOARD FEET OF LUMBER FOR STUD PARTITIONS
2 IN. × 4 IN. STUDS 16 IN. O.C. WITH SINGLE BOTTOM PLATE AND DOUBLE
TOP PLATE

Length of partition wall (ft)	Height of stud partition (ft)					
	$7\frac{1}{2}$	8	$8\frac{1}{2}$	9	$9\frac{1}{2}$	10
5	30	31	33	34	35	37
6	37	39	40	42	44	45
7	44	46	48	50	52	54
8	46	48	50	52	54	56
9	53	55	58	60	62	65
10	60	63	65	68	71	72
11	67	70	73	76	79	82
12	69	72	75	78	81	84
13	76	79	83	86	89	93
14	83	87	90	94	98	101
15	90	94	98	102	106	110
16	92	96	100	104	108	112
17	99	103	108	112	116	121
18	106	111	115	120	125	129
19	113	118	123	128	133	138
20	115	120	125	130	135	140
21	122	127	133	138	143	149
22	129	135	140	146	152	157
23	136	142	148	154	160	166
24	138	144	150	156	162	168
25	145	151	158	164	170	177
26	152	159	165	172	179	185
27	159	166	173	180	187	194

NUMBER OF WOOD JOISTS REQUIRED FOR VARIOUS FLOOR LENGTHS AND
SPACINGS

Joist Spacing (in.)	Distance covered by joists (ft; joists are at right angles to distance)																		
	10	11	12	13	14	15	16	17	18	19	20	21	22	23	24	25	26	27	28
16	9	10	10	11	12	12	13	14	15	16	17	18	18	19	20	21	22	22	
18	8	9	9	10	11	11	12	13	13	14	15	15	16	16	17	17	19	20	
20	7	8	9	9	10	10	11	12	12	13	13	14	15	15	16	16	17	18	18
24	6	7	7	8	8	9	9	10	10	11	11	12	12	13	13	14	14	15	15
26	6	6	7	7	8	8	9	9	10	10	11	11	12	12	13	13	14	14	14
28	6	6	7	7	7	8	8	9	9	10	10	10	11	11	12	12	13	13	13
30	5	6	6	7	7	7	8	8	9	9	9	10	10	11	11	11	12	12	12
32	5	6	6	6	7	7	7	8	8	9	9	9	10	10	10	11	11	12	12
34	5	5	6	6	7	7	7	8	8	8	9	9	9	10	10	11	11	11	
36	5	5	5	6	6	6	7	7	7	8	8	8	9	9	9	10	10	11	

Example: A distance of 18 ft is covered by floor joists 16 in. on center. Therefore, $18 \times 12 = 216$ in. $216 \div 16 = 13.5$ joists, or 14 joists plus $1 = 15$.

Note: A distance of 3 ft at 12-in. spacing requires four joists.
A distance of 4 ft at 12-in. spacing requires five joists.
Therefore, one joist has been added to every distance.

ROUGH CARPENTRY: NAIL CHART

Timber size (in.)	Timber items	Wire nails, common, per 1000 fbm (lb.)		
		10d	20d	30d
6 × 8	Girder stay, bracing, and supports	4	6	
2 × 6	Floor joist	8.5	13	
2 × 8	Floor joist	6.5	9.5	
2 × 10	Floor joist	6.5	10	
2 × 12	Floor joist	6	9	
3 × 8	Floor joist	6	—	12
3 × 10	Floor joist	4.5	—	10
2 × 4	Studs and plates (side wall)	9.5	14.5	
2 × 4	Ceiling beams	6	8.5	
2 × 4	Partition studs, plates, and shoes	11.5	5	
4 × 6	Corner posts	3	6	
2 × 4	Common and hip rafters	15	22	
2 × 6	Common and hip rafters	11	16.5	
2 × 8	Common and hip rafters	8	16.5	
2 × 10	Common and hip rafters	6.5	14.5	
2 × 4	Collar beam	—	24	
2 × 6	Collar beams	—	26	
2 × 8	Collar beams	—	22	
2 × 6	Porch frame and floor joist	3	26	
2 × 8	Porch frame and floor joist	3	26	
2 × 4	Porch ceiling beams	5	11.5	
2 × 6	Porch ceiling beams	7.5	16.5	
2 × 8	Double porch plates	3	26	
2 × 10	Double porch plates	3	26	
2 × 4	Porch rafters	13	32	
2 × 6	Porch rafters	15	32	
2 × 4	Fire blocks	60	—	
2 × 6	Fire blocks	60	—	
2 × 4	Partition bridging	73.5	—	
$1\frac{1}{2}$ × 3	Floor bridging	124	—	

10

FLOORING, TRIM, FLOOR, AND WALL TILE

INTRODUCTION

This unit illustrates and explains the method of estimating wood flooring and trim, ceramic tile, and other types of coverings, such as vinyl, asphalt, rubber, and linoleum.

Wood flooring is made from both hard and soft woods. Flooring made from pine, fir, or other soft woods is commonly 4 and 6 in. wide. Hardwood flooring is made from oak, maple, birch, and beech and is manufactured in a variety of widths and thicknesses. Common face widths of hardwood flooring are $1\frac{1}{2}$, 2, $2\frac{1}{4}$, and $3\frac{1}{4}$ in. Flooring in these widths is manufactured in $\frac{3}{8}$-, $\frac{1}{2}$-, $\frac{5}{8}$-, and $\frac{25}{32}$-in. thicknesses. The $1\frac{1}{2}$ and 2-in. widths are generally available in $\frac{3}{8}$- and $\frac{1}{2}$-in. thicknesses.

Where hardwood flooring is used, oak seems to be the preference of homeowners. It is available in different grades: Clear, Select, No. 1 Common, and No. 2 Common. The No. 1 Common grade is used more extensively than other grades.

Clear is practically free of defects. Bright sap marks up to an average width of about $\frac{3}{8}$ in. are permitted. Select may contain sap, pinworm holes, streaks, slight imperfections in working, and small, tight knots not to exceed one in every 3 ft. of length.

No. 1 Common contains more imperfections than Select, such as streaks, checks, and knots, but it makes a good, sound floor.

No. 2 Common contains still more imperfections than No. 1 Common.

Oak flooring boards are bound in bundles.   Each bundle is marked showing the average length of pieces in the bundle, such as 2, 3, 4, 5, 6 ft, etc.   The number of boards or pieces in a bundle vary from 24, 12, and 8, depending on size (thickness and width) of the flooring.   There is considerable loss of surface due to milling to finish size and a little loss or waste in cutting on the job.   In estimating, therefore, it is necessary to add allowances to the net floor area, as given in Fig. 10-1.

THE TAKE-OFF

When estimating wood flooring, find the square feet of surface to be covered and then add the percentages given in Fig. 10-1 for the size of the flooring to be used. A carpenter will take about 28 hr (first-class work) to lay 1000 B.F. of $\frac{23}{32}$ in. $\times$ $2\frac{1}{4}$ in. oak flooring.

For example, figure the board feet of No. 1 Common oak flooring for the living room, dining room, and bedroom (closets included) in the plan in Fig. 10-2.   The flooring board size used is $\frac{23}{32}$ in. $\times$ $2\frac{1}{4}$ in.

ALLOWANCES TO BE ADDED FOR ESTIMATING QUANTITIES OF HARDWOOD FLOORING	
Nominal Sizes (Actual Face Widths) (Inches)	Allowances (%)
$\frac{3}{8}$ x $1\frac{1}{2}$	35
$\frac{3}{8}$ x 2	27
$\frac{1}{2}$ x $1\frac{1}{2}$	35
$\frac{1}{2}$ x 2	27
$\frac{25}{32}$ x $1\frac{1}{2}$	52
$\frac{25}{32}$ x 2	40
$\frac{25}{32}$ x $2\frac{1}{4}$	35
$\frac{25}{32}$ x $3\frac{1}{4}$	25

Figure 10-1

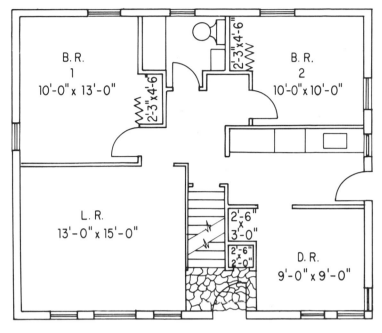

Figure 10-2

			Square feet
Living room	13 ft × 15 ft	=	195
Hall closet	2'0" × 2'6"	=	5
Dining room	9 ft × 9 ft	=	81
Alcove	2'6" × 3'0"	=	7.5
Bedroom 1	10 ft × 13 ft	=	130
Closet	2'3" × 4'6"	=	10.125
Bedroom 2	10 ft × 10 ft	=	100
Closet	2'3" × 4'6"	=	10.125

538.75, or 539, sq ft

Then

$$539 \text{ sq ft} \times 1.35 = 728 \text{ sq ft of flooring}$$

In figuring board feet of flooring, $\frac{25}{32}$ in. is considered 1 in. Therefore, 728 sq ft of this flooring is 728 B.F. If the laying of 1000 B.F. requires 28 hr of labor, 1 B.F. requires $\frac{28}{1000}$, or $28 \div 1000 = 0.028$ hr/B.F. Then

$$0.028 \text{ hr/B.F.} \times 728 \text{ B.F.} = 20.384, \text{ or } 21, \text{ hr of labor time}$$

Prefinished block flooring is furnished in squares usually 9 in. × 9 in. with a thickness of $\frac{25}{32}$ in. They are made from hardwoods such as oak, beech, maple, and walnut. These blocks may be laid in mastic over concrete, plywood, or similar subfloors. They can also be laid over old wood floors.

It takes 100 sq ft of blocks to cover 100 sq ft of flooring. There are sixteen 9 in. × 9 in. blocks to the carton, which cover an area of 9 sq ft. To estimate the number of cartons, find the area of the floor and divide by 9 sq ft.

Example

A floor is 12 ft × 18 ft. How many cartons of 9 in. × 9 in. block flooring are required?

Solution

$$12 \text{ ft} \times 18 \text{ ft} = 216 \text{ sq ft}$$

$$216 \text{ sq ft} \div 9 \text{ sq ft} = 24 \text{ cartons}$$

Determining the Amount of Floor Felt

Usually, floor felt is desirable under oak flooring. This is figured by getting the square feet of flooring and adding 10%.

In Fig. 10-2 we found 539 sq ft for all rooms. Therefore,

$$539 \text{ sq ft} + 53.9 \text{ sq ft} (10\%) = 592.9, \text{ or } 593, \text{ sq ft of felt}$$

Figuring the Base Molding

The perimeter of each room where base molding is required is measured. Thus, if a room measures 10 ft × 12 ft its perimeter is 44 lin ft. Where many doors or large openings occur, some deduction should be made. The perimeter of all rooms is added, and the amount of base mold is thus determined. The final answer is in linear feet.

Figuring the Amount of Ceramic Tile

Floor and wall tile are measured in square feet. When a floor or a wall is irregular in shape, it can be divided into squares or rectangles. Then the areas are calculated to find the total area of the floor or wall. Caps and base tile are figured in linear feet. Waste and breakage is figured by allowing 1 to 4 sq ft of tile for the floor and about the same for the wall tile (depending a lot on the amount of cutting of tile required for a particular job). Also, allowance should be made of about 2 lin ft each for caps and base tile.

A floor tile setter, including a helper, can set about 8 sq ft/hr. For wall tile, approximately 7 sq ft/hr can be set.

Problem

Figure the square feet of ceramic tile 4¼ in. × 4¼ in. required for the floor and walls of the bathroom shown in the plan and elevations in Fig. 10-3. Also, figure the linear feet of caps 2 in. × 6 in. and the base tile 4 in. × 6 in.

Solution   Floor tile:

$$9 \text{ ft} \times 5.5 \text{ ft} = 49.5, \text{ or } 50 \text{ sq ft (plan)}$$

Allow 2 sq ft for waste = 52 sq ft of floor tile.

Wall tile:

$$\text{(Above tub) } 2.5 \text{ ft} + 5 \text{ ft} + 2.5 \text{ ft} = 10 \text{ ft (section A-A)}$$

$$10 \text{ ft} \times 5 \text{ ft (height)} = 50 \text{ sq ft}$$

$$\text{(Other walls) } 3 \text{ ft} + 5.5 \text{ ft} + 9 \text{ ft} + 2.08 \text{ ft} + 0.5 \text{ ft} + 1 \text{ ft} = 21.08 \text{ ft}$$

$$21.08 \text{ ft} \times 2.83 \text{ ft (height)} = 59.65, \text{ or } 60, \text{ sq ft}$$

$$50 \text{ sq ft} + 60 \text{ sq ft} = 110 \text{ sq ft}$$

Allow 3 sq ft for waste = 113 sq ft of wall tile.

Base tile:

$$21.08 \text{ ft} + \text{ about 2 ft waste} = 23 \text{ lin ft of base tile}$$

Tile caps:

$$10 \text{ ft. (above tub)} + 21.08 \text{ ft} = 31.08 \text{ lin ft}$$

Allow about 2 ft for waste = 33 lin ft of tile caps

Problem

Find the vinyl floor covering required for the floor in Fig. 10-4. The material to be purchased is 6′0″ wide.

Solution   The rectangles in Fig. 10-4 show the plan for laying the material on the floor. It will require two strips 6 ft wide, each strip being 15 ft long. The waste strip is 6 in. wide and 15 ft long. Therefore, the amount of 6 ft material required will be

$$2 \times 15 \text{ ft} = 30 \text{ ft}$$

$$6 \text{ ft} \times 30 \text{ ft} = 180 \text{ sq ft}$$

$$180 \text{ sq ft} \div 9 \text{ sq ft} = 20 \text{ sq yd of vinyl required}$$

Vinyl, Rubber, and Asphalt Floor Tile

This material is furnished in cartons containing 80 pc. of 9 in. × 9 in. tile. One carton covers 45 sq ft of surface. If felt lining is used when applied on wood floors, an adhesive is brushed on the floor before the felt is applied. On top of the felt lining a paste is troweled before the tile is set.

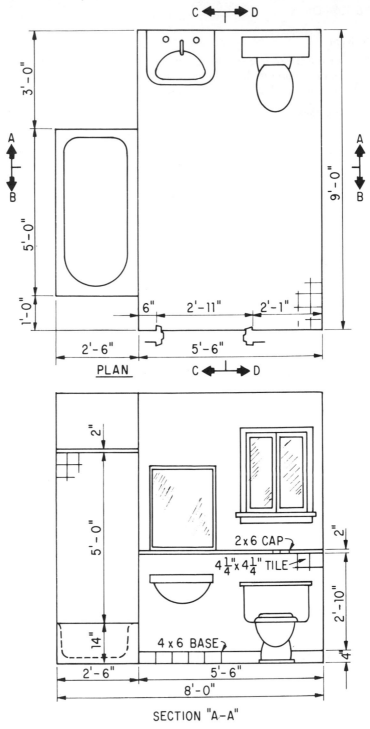

PLAN

SECTION "A-A"

Figure 10-3

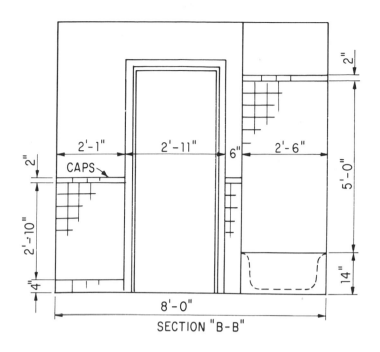

SECTION "B-B"

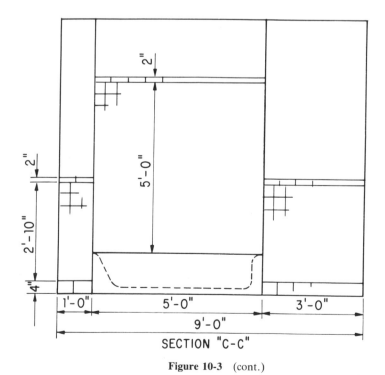

SECTION "C-C"

Figure 10-3 (cont.)

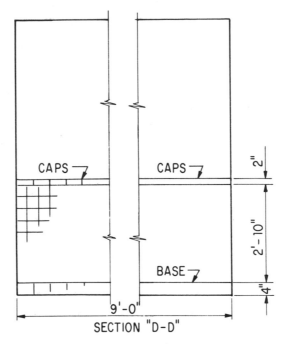

CAPS

CAPS

2"

2'-10"

BASE

4"

9'-0"

SECTION "D-D"

Figure 10-3 (cont.)

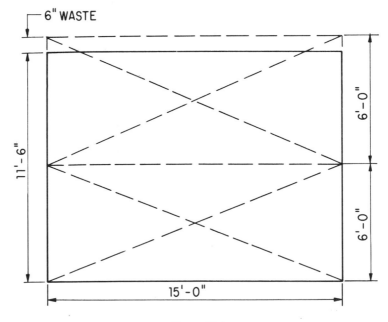

6" WASTE

11'-6"

6'-0"

6'-0"

15'-0"

Figure 10-4

On concrete floors the tile is usually laid directly on the concrete after the paste is applied.

In estimating 9 in. × 9 in. floor tile, find the area to be covered and divide by 45 to get the number of cartons.

Problem

A floor measures 10 ft × 18 ft. Find the number of cartons of asphalt floor tile required when 9 in. × 9 in. tile is used

Solution   The area of room is 10 ft × 18 ft = 180 sq ft

$$180 \text{ sq ft} \div 45 \text{ sq ft} = 4 \text{ cartons}$$

Allow for cutting and waste.

MATERIAL AND INSTALLATION COSTS

Wood floor—maple: vertical grain $\frac{25}{32} \times 2\frac{1}{4}$, select, no finish

$1.80 per square foot for material (and $0.04 per square foot for underlayment)
 0.90 per square foot for installation
$2.70 per square foot

Including subcontractor's overhead and profit, $3.24 per square foot.

Wood floor—oak: $\frac{25}{32}$ in. × $2\frac{1}{4}$, no finish, No. 1 common

$1.20 per square foot for material (add 4¢ per square foot for underlayment)
 0.81 per square foot for installation
$2.01 per square foot

Including subcontractor's overhead and profit, $2.41 per square foot.

Wood floor—yellow pine: $\frac{25}{32}$ in. × $3\frac{1}{4}$ in. T&G grade C, no finish included

$1.20 per square foot for material
 0.74 per square foot for installation
$1.94 per square foot

Including subcontractor's overhead and profit, $2.38 per square foot.

Vinyl asbestos tile: 12 in. × 12 in. × $\frac{1}{16}$ in. thick

$0.52 per square foot for material
 0.28 per square foot for installation
$0.80 per square foot

Including subcontractor's overhead and profit, $0.96 per square foot.

Ceramic tile: walls, interior, adhesive set $4\frac{1}{4}$ in. $\times$ $4\frac{1}{4}$ in. tile

$1.42 per square foot for material
  1.43 per square foot for installation
$2.85 per square foot

Including subcontractor's overhead and profit, $3.53 per square foot.
   Asphalt tile: on concrete $\frac{1}{8}$ in. thick

$0.33 per square foot for material (add $0.04 per square foot for underlayment)
 0.20 per square foot for installation
$0.53 per square foot

Including subcontractor's overhead and profit, $0.64 per square foot.

SELF EXAMINATION

Find the amount of 6-ft-width vinyl required for the kitchen shown in Fig. 10-5.

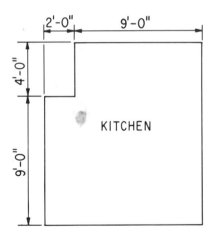

Figure 10-5

Answer Box

The most economical piece of vinyl will be 6'0" wide by 23'0" long

---

ASSIGNMENT—10

1. Find the board feet of oak flooring required for the bedrooms (including closets) and the dining and living rooms in the plan in Fig. 10-6.   The board size is to be $\frac{25}{32}$ in. $\times$ $2\frac{1}{4}$ in. Include material and installation costs.
2. Figure the amount of floor felt needed for the flooring in Problem 1.

3. Figure the amount of one-piece base molding for the bedrooms (including closets) and the dining and living rooms (Fig. 10-6). Make no allowance for the door openings (extra material is usually allowed to avoid unnecessary piecing).

4. Figure the square feet of vinyl asbestos tile 12 in. × 12 in. required for the kitchen floor shown in the plan in Fig. 10-6. Exclude the area under the cabinets. Allow 10% for waste. Add material and installation costs.

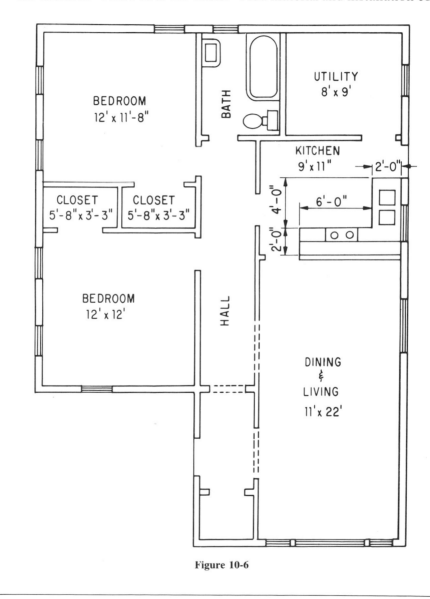

Figure 10-6

SUPPLEMENTARY INFORMATION

AMOUNT OF SURFACE 1000 FT OF FLOORING WILL COVER AND
QUANTITY OF NAILS REQUIRED TO LAY IT

Nominal size (in.)	Actual size (in.)	Square feet of floor covered	Nail spacing (in.)	Nails required (lb)	Type of Nails
1×2	$\frac{3}{4} \times 4\frac{1}{2}$	750	8	20	4d Casing
$1 \times 2\frac{1}{4}$	$\frac{13}{16} \times 1\frac{1}{2}$	667	12	70	8d Coated casing
$1 \times 2\frac{1}{2}$	$\frac{3}{8} \times 2$	800	8	17	4d Casing
$1 \times 2\frac{3}{4}$	$\frac{13}{16} \times 2$	727	12	56	8d Coated casing
1×3	$\frac{13}{16} \times 2\frac{1}{4}$	750	10	64	8d Coated casing
1×4	$\frac{13}{16} \times 3\frac{1}{4}$	800	10	29	8d Coated casing

REVIEW QUESTIONS

1. Hardwood flooring is made from what woods?

2. Flooring of $\frac{25}{32}$ in. $\times$ $1\frac{1}{2}$ in. requires an additional allowance of what percentage?

3. A wall tile setter can set how many square feet per 8-hr day?

4. Vinyl tile of 9 in. $\times$ 9 in. size contains how many pieces in a carton?

11

LATHING AND PLASTERING

INTRODUCTION

In this unit you will learn how to arrive at the various quantities of materials required for lathing and plastering.

The word "lathing" suggests the material that is first applied to wall studs or ceiling beams before the wet plaster is applied. The lathing gives sufficient keying or bonding of the plaster.

There are three types of lath: wire lath, plain gypsum lath, and perforated gypsum lath. All three are intended as the bounding base for the wet plaster (see Fig. 11-1).

Wire lath, often known as "diagonal" metal lath, is packed in bundles of 10 sheets, each sheet 27 in. wide and 96 in. long. A 27 in. (2.25 ft) × 96 in. (8 ft) = 18 sq ft. Ten sheets × 18 sq ft = 180 sq ft per bundle, or 180 ÷ 9 sq ft = 20 sq yd per bundle.

Wire lath is generally specified in weights of 2.5 or 3.4 lb per square yard, black-dip-painted finish, or it may be galvanized coated.

To prevent some of the wet plaster from squeezing through the wire lath and falling behind the studs as waste, an asphalt-saturated building paper of 15 lb per 100 sq ft, is first applied to the wall studs.

Exterior stucco walls require a stucco mesh, also known as chicken wire. When used as a base material for exterior cement plaster, it is installed over 15-lb waterproof building paper, fastened to the wall with double-head furring nails.

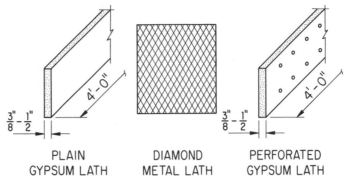

PLAIN | DIAMOND | PERFORATED
GYPSUM LATH | METAL LATH | GYPSUM LATH

Figure 11-1

Plain gypsum lath is furnished in panels of 16 in. × 4 ft × $\frac{3}{8}$ or $\frac{1}{2}$ in. thick. The plaster core is sandwiched between two layers of heavy paper. One bundle of six lath units covers 32 sq ft. Ten square yards per hour can be installed on walls in an 8-hr day.

Perforated gypsum lath is furnished in panels of 16 in. × 4 ft × $\frac{2}{8}$ or $\frac{1}{2}$ in. thick. The core is of plaster housed between two layers of heavy paper. This lath has $\frac{3}{4}$-in. holes, 4 in. o.c. aligned both vertically and horizontally. The holes are the keys for the wet plaster.

Plastering refers to the application of the various types of plaster to a lathed surface in layers to give a smooth or textured hard surface finish ready to receive paint or other surface finishes.

For exterior plastering, portland cement-type plaster is used; and for interior plastering, gypsum plaster is generally used.

THE TAKE-OFF

Estimating the Amount of Metal Lath

In estimating the amount of metal lath required for a room, multiply the total linear feet of the walls around the room by the height of the wall to get the total number of square feet. Do not deduct for windows and doors (this extra amount will replace the waste in cutting). However, where very large openings exist, such as window walls or large picture windows, such areas should be deducted. For ceilings, multiply the length by the width. When the square feet of a room to be covered by metal lath are known, divide the square feet by 9 to get the number of square yards.

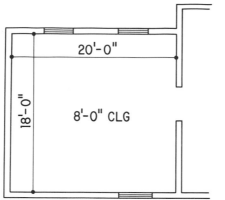

Figure 11-2

To find the number of metal lath bundles, proceed as follows. A 10-sheet bundle of 27 in. × 96 in. contains 20 sq yd. Divide the total square yards of the room by 20 to get the number of bundles.

For example, find the bundles of metal lath (10 sheets to the bundle) for the room shown in Fig. 11-2

Area of walls: 20 ft + 20 ft + 18 ft + 18 ft
= 76 lin ft × 8 ft (ceiling height) = 608 sq ft

Area of ceiling: 18 × 20 ft = 360 sq ft

Total square feet of room: 608 sq ft + 360 sq ft = 968 sq ft

968 sq ft ÷ 9 sq ft = 107.5 sq yd

107.5 sq yd ÷ 20 sq yd = 5.375, or 6, bundles required

The preceding example can be stated simply in tabular form:

METAL LATH: 10-SHEET BUNDLES AT 20 SQ YD/BUNDLE

Item	Unit	Length (ft)	Width (ft)	Height (ft)	Square feet	Square yards	Bundles
Wall	1	76		8	608		
Ceiling	1	20	18		360		
					968	107.5	6

Problem

In tabular form take-off the quantities of metal lath required for the rooms shown in the plan, in Fig. 11-3. Find the number of 10-sheet bundles of 27 in. × 96 in. sheets.

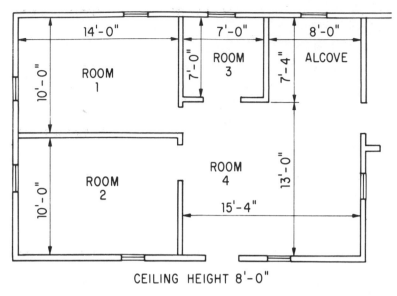

Figure 11-3

Solution

METAL LATH: 10-SHEET BUNDLES AT 20 SQ YD/BUNDLE

Item	Unit	Length (ft)	Width (ft)	Height (ft)	Square feet	Square yards	Bundles
Rooms 1 and 2							
Walls	2	48		8	768		
Ceilings	2	14	10		280		
Room 3							
Walls	1	28		8	224		
Ceiling	1	7	7		49		
Alcove							
Walls	1	22.67		8	181.36		
Ceiling	1	9	7.33		58.64		
Room 4							
Walls	1	48.67		8	389.36		
Ceiling	1	15.33	13.00		199.29		
					2149.65	239.0	12

Nails for metal lath.   The general practice is to allow 2 lb of $1\frac{1}{2}$-in. or $1\frac{1}{4}$-in. roofing nails per 10-sheet bundle of metal lath.

Estimating Gypsum Lath

To estimate gypsum lath, find the total area of the surfaces, walls, and ceilings and deduct all window and door openings. Divide the square-foot area of one panel into the area of the surface to be covered. The result is the number of panels of the size selected.

Estimating Corner Beads

Corner beads (Fig. 11-4) are used around openings and at corners of plastered walls to prevent chipping. Stock length of corner beads are 8, 9, 10, and 12 ft.

    To estimate corner beads, find the combined length of outside corners in all rooms to be plastered. If any two rooms are connected by a plastered archway, multiply the distance around the archway by 2 and add this number of linear feet of corner beads to any previously found.

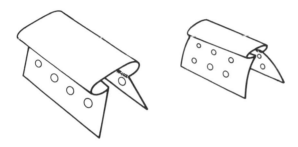

TYPICAL CORNER BEADS   **Figure 11-4**

Problem

    Estimate the linear feet of corner beads required around the arch openings and plaster corner shown in the plan in Fig. 11-5. Hall arch openings are 3'0" × 7'0". The kitchen–dining room arch is 2'8" × 7'0". The ceiling height is 8'0".

Solution The corner bead at A is 8'0". The distance around arch B is 7 ft + 3 ft + 7 ft = 17 ft × 2 (two sides) = 34 ft

$$34 \text{ ft} \times 2 \text{ (for two arches)} = 68 \text{ ft}$$

The distance around arch C is 7 ft + 2.67 ft + 7 ft = 16.67 ft

$$16.67 \text{ ft} \times 2 \text{ (two sides)} = 33.34, \text{ or } 34, \text{ ft}$$

The total length is

$$8 \text{ ft} + 68 \text{ ft} + 34 \text{ ft} = 110 \text{ lin ft of corner beads}$$

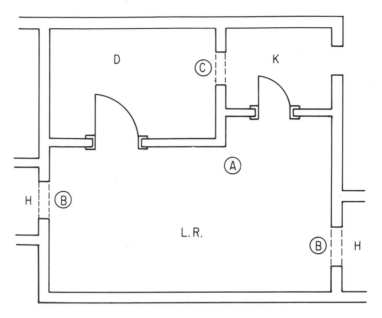

Figure 11-5

Estimating Gypsum Wallboard

When ordering corner beads, select a suitable stock length and divide the combined length of bead required by the stock length desired, counting any fractional part of a piece as one additional length.

Gypsum wallboard is a mill-fabricated wallboard composed essentially of a fireproof gypsum core encased in a heavy, manila-finished paper on the face side and a strong, liner paper on the back side. Gypsum wallboard takes the place of a plaster wall or ceiling. The panels, furnished 4 ft wide by 8, 9, 10, 12, and 14 ft long, are nailed to the inside stud faces and ceiling joists. Where studs are spaced 12 and 16 in. on centers, $\frac{3}{8}$- and $\frac{1}{2}$-in.-thick panels are satisfactory. On stud spacings 24 in. on centers, the panels should be $\frac{1}{2}$ or $\frac{5}{8}$ in. in thickness. Gypsum wallboard $\frac{1}{4}$ in. thick may be used on curved surfaces or for refacing an old wall.

In estimating gypsum wallboard, find the area of the walls and ceilings to be covered. Deduct all areas for windows and doors. Divide the total net square feet by the square feet for one panel in order to find the total number of panels. Since there is practically no waste in the application of the panel, no extra allowance need be made. However, a full additional panel is ordered when computations result in fractions of a panel.

Problem

The walls and ceiling of a room 10 ft. × 18 ft. × 8 ft. high are to be paneled with gypsum wallboard. The panel width is 4 ft. × 7′0″. Find the number of panels required.

Solution

$$\text{Ceiling: } 18 \text{ ft} \times 10 \text{ ft} = 180 \text{ sq ft}$$

$$\text{Walls: } 10 \text{ ft} + 18 \text{ ft} + 10 \text{ ft} + 18 \text{ ft}$$

$$= 56 \text{ lin ft} \times 8 = \underline{448 \text{ sq ft}}$$
$$628 \text{ sq ft}$$

$$\text{Windows: } 2 \times 2.67 \text{ ft} \times 4.17 \text{ ft} = 22.26 \text{ sq ft}$$

$$\text{Door: } 1 \times 3 \text{ ft} \times 7 \text{ ft} = \underline{21.00 \text{ sq ft}}$$
$$43.26 \text{ sq ft}$$

Then

$$628 \text{ sq ft} - 43.26 \text{ sq ft} = 584.74, \text{ or } 585, \text{ sq ft net}$$

$$\text{Panel size: } 4 \text{ ft} \times 8 \text{ ft} = 32 \text{ sq ft}$$

$$585 \text{ sq ft} \div 32 \text{ sq ft} - 18.28, \text{ or } 19, \text{ panels}$$

Estimating the Quantities of Plaster

All plaster work is measured by the square yard for flat surfaces. Find the areas of all walls and ceilings to be plastered less 7% of the area for doors and window openings. Large openings, however, such as window walls or large picture windows should be deducted from the total area before the 7% deduction is made for smaller openings and waste.

For example, in determining the number of square yards of plastering in a room 18 ft × 20 ft having an 8-ft ceiling height, take the perimeter of the room, such as

$$18 \text{ ft} + 20 \text{ ft} + 18 \text{ ft} + 20 \text{ ft} = 76 \text{ lin ft of wall}$$

$$76 \text{ ft} \times 8 \text{ ft (ceiling height)} = 608 \text{ sq ft of wall}$$

$$18 \text{ ft} \times 20 \text{ ft} = 360 \text{ sq ft of ceiling}$$

The total area, therefore, is 608 sq ft + 360 sq ft = 968 sq ft

$$968 \text{ sq ft} \div 9 \text{ sq ft} = 107.5, \text{ or } 108 \text{ sq yd}$$

$$7\% \text{ of } 108 \text{ sq yd} = 0.07 \times 108 \text{ sq yd} = 7.56 \text{ sq yd}$$

$$108 \text{ sq yd} - 7.56 \text{ sq yd} = 100.44 \text{ sq yd}$$

Plastering. All plaster contains at least one cementitious substance, such as lime, gypsum, or portland cement. Most plaster is prepared by mixing one or two of these cementitious substances with sand or other aggregates such as perlite or vermiculite, according to specifications.

Lime, gypsum, and cement plaster.   A distinction must be made among lime, gypsum, and cement plaster.   While gypsum plaster is used on gypsum blocks and plasterboards, cement plaster is used in places where fire resistance or water resistance is required.   Although lime was formerly used alone with sand in preparing plaster, it is now more commonly used along with other cementitious materials.   The first plaster coat, called the *scratch coat*, consists of gypsum and sand and is forced into the metal lath so that some of the material comes through and forms a key.   The lath is covered to a $\frac{1}{4}$-in. thickness.   Before it hardens it is scratched with a tool or a piece of metal lath so that a rough surface is created that will act as a base for the next coat.   The second coat, called the *brown coat*, is the thickest of the three coats.   It should be spread to a $\frac{3}{8}$- or $\frac{1}{2}$-in. thickness. It consists of gypsum cement plaster and sand and is troweled to a true surface with a darby or straightedge and finished with a float.   The third coat, called the *finish coat*, consists of lime and gauging plaster in a thickness of about $\frac{1}{8}$ in., and it is troweled with a steel trowel to a hard, smooth finish ready for painting.

Material quantity for scratch coat.   The scratch coat for metal lath generally requires ten 100-lb bags of gypsum cement plaster and 1 yd of sand per 100 sq yd of surface.   The scratch coat for gypsum lath requires five 100-lb bags of gypsum cement plaster and $\frac{1}{2}$ yd of sand per 100 sq yd of surface.

Material quantity for brown coat.   The brown coat is usually in the proportion of 1 part plaster to 3 parts sand.   For 100 sq yd of application, seven bags of gypsum cement plaster and 21 cu ft of sand are necessary.

100 SQ. YD. OF PLASTER FOR VARIOUS APPLICATIONS			
Type of Application	Gypsum Cement Plaster	Sand	Surface
Scratch coat	10 bags	1 yd.	Metal lath
Scratch coat	5 bags	1/2 yd.	Gypsum lath
Brown coat	7 bags	21 cu. ft.	Scratch coat
Finish coat	7 bags lime	150 lb. gauging plaster	Brown coat

Figure 11-6

Material quantity for finish coat.   The finish coat is in the proportion of one bag (50 lb) of lime and $21\frac{1}{2}$ lb of gauging plaster.   For 100 sq yd of application, seven bags of lime and 150 lb of gauging plaster are required.

The quantities of plaster per 100 sq yd for the various applications may simply be stated in tabular form, as in Fig. 11-6.

Problem

In the plan in Fig. 11-7, estimate the following quantities of materials: (1) scratch coat on metal lath, (2) brown coat on scratch, and (3) finish coat on brown.

Solution   The combined inside length of walls for the room is

$$\text{Walls: } 25 \text{ ft} + 18 \text{ ft} + 25 \text{ ft} + 18 \text{ ft} = 86 \text{ lin ft} \times 8 \text{ ft}$$
$$= 688 \text{ sq ft}$$

$$\text{Ceiling: } 25 \text{ ft} \times 18 \text{ ft} = 450 \text{ sq ft}$$

$$688 \text{ sq ft} + 450 \text{ sq ft} = 1138 \text{ sq ft}$$

$$1138 \text{ sq ft} \div 9 \text{ sq ft} = 126.4, \text{ or } 127 \text{ sq yd}$$

$$7\% \text{ of } 127 \text{ sq yd} = 0.07 \times 127 = 8.89 \text{ sq yd}$$

$$127 \text{ sq yd} - 8.89 \text{ sq yd} = 118.11 \text{ sq yd}$$

$$118.11 \text{ sq yd} \div 100 = 1.18 \text{ hundreds of square yards}$$
$$\text{to be plastered}$$

Therefore,

$$\text{Scratch coat (see Fig. 11-6) 10 bags} \times 1.18$$
$$= 11.8, \text{ or } 12 \text{ bags of gypsum cement plaster}$$

$$1 \times 1.18 = 1.18 \text{ cu yd of sand}$$

$$\text{Brown coat: 7 bags} \times 1.18 = 8.26 \text{ or } 9 \text{ bags of gypsum cement plaster}$$

$$21 \text{ cu ft of sand} \times 1.18 = 24.78, \text{ or } 25 \text{ cu ft of sand}$$

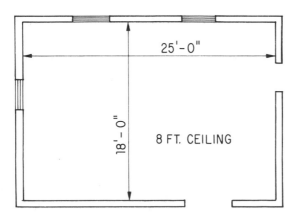

25'-0"

18'-0"

8 FT. CEILING

Figure 11-7

Finish coat: 7 bags of lime $\times$ 1.18 = 8.26, or 9 bags of lime

150 lb of plaster $\times$ 1.18 = 177 lb of gauging plaster

The preceding problem can be stated simply and solved in tabular form as shown in Fig. 11-8.

Item	Unit	Length (ft.)	Height (ft.)	Width (ft.)	Square Feet	Square Yards	Scratch Coat 10 Bags Gypsum Plaster, 1 cu. yd. Sand	Brown Coat 7 Bags Gypsum Plaster, 21 cu. ft. Sand	Finish Coat 7 Bags Lime, 150 lb. Gauging Plaster
Walls	1	86	8		688				
Ceiling	1	25		18	450				
					1138	127 7% 118 sq. yds. 1.18 Hundreds of Sqare Yards	1.18 x 10 12 Bags Gypsum Plaster 1.18 x 1 = 1.18 cu. yds. Sand	1.18 x 7 9 Bags Gypsum Plaster 1.18 x 21 = 25 cu. yds. Sand	1.18 x 7 9 Bags Lime 1.18 x 150 = 177 lb. Gauging Plaster

Figure 11-8

MATERIAL AND INSTALLATION COSTS

Wire lath—diamond (2.5 lb): painted, on wood framing walls

$1.53 per square yard for material
 1.76 per square yard for installation
$3.29 per square yard

With subcontractor's overhead and profit, $4.11 per square yard.   Lather at $18.95 per hr can install 85 S.Y.   per 8-hr day.

Wire lath—diamond (3.4 lb): painted, on wood framing walls

$1.84 per square yard for material
 1.89 per square yard for installation
$3.73 per square yard

With subcontractor's overhead and profit, $4.64 per square yard.   Lather at $18.95 per hour can install 80 S.Y. per 8-hr day.

Corner bead: galvanized $1\frac{1}{4}$ in. $\times$ $1\frac{1}{4}$ in.

> $ 9.35 per hundred linear feet for material
> $\underline{54.00}$ per hundred linear feet for installation
> $63.35 per hundred linear feet

With subcontractor's overhead and profit, $79.00 per hundred linear feet.

Gypsum plaster board: $\frac{3}{8}$ in. thick nailed to studs, no finish

> $0.18 per square foot for material
> $\underline{0.14}$ per square foot for installation
> $0.32 per square foot

With subcontractor's overhead and profit, $0.37 per square foot. For topping and finishing joints add $0.01 per square foot for material add $0.11 per square foot for installation

Gypsum lath: with two coats vermiculite plaster, two sides, on 2 in. $\times$ 4 in. wood studs, 16 in. o.c.

> $1.28 per square foot for material
> $\underline{2.50}$ per square foot for installation
> $3.78 per square foot

With subcontractor's overhead and profit, $4.72 per square foot.

Plaster, perlite, or vermiculite: two coats, no lath included, on walls

> $2.49 per square yard for material
> $\underline{6.81}$ per square yard for installation
> $9.30 per square yard

With subcontractor's overhead and profit, $12.10 per square yard and ceiling, $12.45 per square yard

Three plasterers and one lather can install 87 sq yd of a three-coat plaster job on walls in an 8-hr day—on ceilings 78 sq yd. 800 sq ft of gypsum wall board, 48 in. $\times$ 96 in. $\times$ $\frac{1}{2}$ in. or $\frac{5}{8}$ in. can be installed in an 8-hr day. The joints between gypsum wall board are covered with perforated paper tape and installed with joint compound. For 1000 sq ft of area, 400 lin ft of tape and 75 lb of prepared compound is required. About 250 lin ft of wire-tied corner beads can be installed in an 8-hr day.

SELF EXAMINATION

Find the number of metal lath bundles and the quantities of plaster required for the two bedrooms and kitchen only (excluding closets) in Fig. 11-9. The ceiling height is 8 ft. Metal lath: Use a 10-sheet bundle, 27 in. $\times$ 96 in., containing 20 sq yd. Plaster: scratch coat, brown coat, and finish coat.

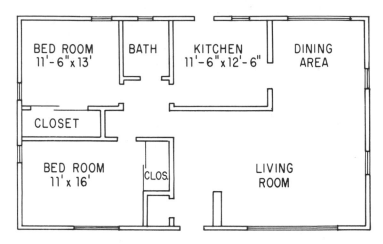

Figure 11-9

Answer Box	
Metal lath	9.35, or 10, bundles
Plaster: Scratch coat	18 Bags gypsum cement plaster + 2 yd sand
Brown coat	13 Bags gypsum cement plaster + 37 cu ft sand
Finish coat	13 Bags gypsum cement plaster + 261 lb of gauging plaster

ASSIGNMENT—11

1. Find the number of metal lath bundles and the quantity of plaster required for the living room, dining room, kitchen, and bedroom only (exclude closet) for the plan shown in Fig. 11-10. The ceiling height is 8 ft. Metal lath: Use a 10-sheet bundle 27 in. × 96 in. (20 sq yd to the bundle). Plaster: scratch coat, brown coat, and finish coat. *Note:* Take-off quantities in tabular form.
2. Find the material and installation costs of the metal lath.
3. Total cost of plaster work for both ceiling and walls. Costs with overhead and profit.

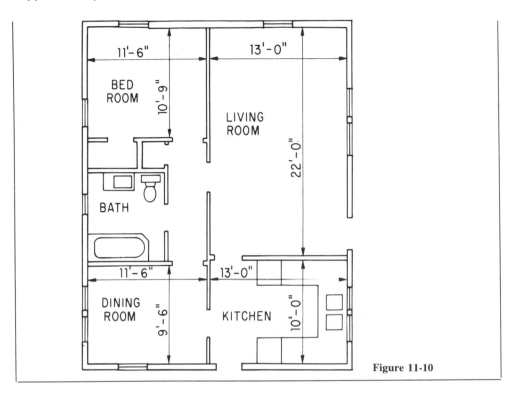

Figure 11-10

SUPPLEMENTARY INFORMATION

Data on Threaded Nails

Two types of nails, the helically grooved and the annularly grooved, have recently been mass produced so that they can compete in price with the plain shank nail (see Fig. 11-11).

The helically grooved nail has a long pitch shank that permits it to screw its way into the wood. The nail displaces the fibers and forms a thread in the wood. This compresses the surrounding fibers and increases the frictional resistance between the wood and the shank, and thus the holding power of the nail.

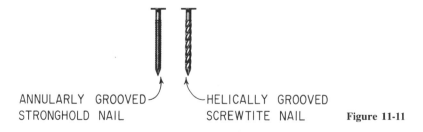

ANNULARLY GROOVED
STRONGHOLD NAIL

HELICALLY GROOVED
SCREWTITE NAIL

Figure 11-11

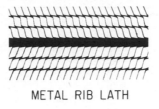

METAL RIB LATH

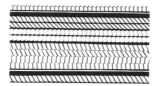

FLAT RIB METAL LATH   **Figure 11-12**

The annularly grooved nail has numerous grooves encircling the shank and when driven into the wood forces the fibers into the annular groovelike wedges to be released only when the wood is destroyed.

Careful and extensive research has shown that helically grooved and annularly grooved nails, although harder to drive into wood, offer a greater load capacity and are harder to withdraw.

Figure 11-12 shows two kinds of metal lath used as a plaster base: The rib lath and the flat rib lath.

Figure 11-13 illustrates metal lath nailed to studs and acting as a key for the scratch coat.

Figure 11-14 illustrates a scratch coat applied to perforated gypsum lath.

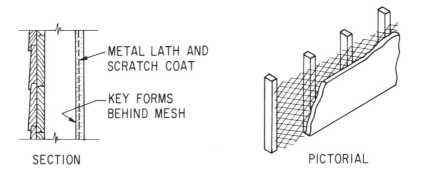

METAL LATH AND
SCRATCH COAT

KEY FORMS
BEHIND MESH

SECTION   PICTORIAL

WIRE LATH AND PLASTER

Figure 11-13

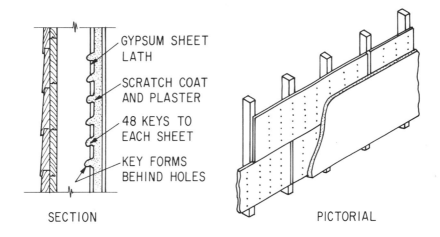

SECTION   PICTORIAL

SHEET LATH AND PLASTER

Figure 11-14

REVIEW QUESTIONS

1. How is diamond metal lath packaged?
2. Explain how metal lath is estimated.
3. What is gypsum lath?
4. How is gypsum plaster board estimated?
5. A 10 ft × 12 ft room with an 8-ft ceiling height requires how many gypsum plaster boards for walls and ceiling?   What is the cost for material and installation?   The cost with a 30% markup?

12

ROOFING
AND WALL SHINGLES

INTRODUCTION

Shingles are manufactured in various materials such as wood, slate, clay tile, asbestos-cement, asphalt, aluminum, and special coated steel.

Wood shingles are single shingles and are manufactured in three types: dimension, random, and hand split. They are installed either on solid sheathing or on wood strips. The exposure varies with the roof pitch.

Slate may be laid on wood strips spaced the same as wood exposure or on solid sheathing with 15-lb felt first applied.

Clay tile shingles can only be used on roofs with pitches $4\frac{1}{2}$ to 12 and greater. These shingles are made from the same clay as used for bricks and are made in types known as Spanish, Mission, French, English, Norman, and Scandia.

Asbestos-cement shingles are made from portland cement and asbestos fibers, which act as reinforcing. They are available either plain or striated or with a good-textured surface, and in a variety of colors.

Slate, clay tile, and asbestos-cement shingles are very heavy material and therefore the supporting structure must be designed to support this extra load. Asbestos-cement shingles vary in weight from 265 to 585 lb per square (100 sq ft) slate weighs 700 to 810 lb per square, and clay-tile weighs 900 lb per square.

Aluminum and steel shingles are available with colored baked enamel, vinyl, or special organic coatings. These shingles are similar to clay-tile shingles in form and shape but are much lighter.

In this unit we will deal with the estimating of wood and asphalt shingles required on a house, together with the extra quantities required for the doubling of the starter shingle courses.

Wood Shingles

Wood shingles are laid on exterior walls in several ways. The common method is *single coursing*. Each single course is uniformly spaced or varied slightly to bring about alignment of the bottom of the courses with the tops and bottoms of openings.

Another popular method of laying shingles on walls is called *double coursing*. By this method two thicknesses of shingles are laid in each course with the butts of the outer course extending approximately $\frac{1}{2}$ in. below the inner or concealed course (see Fig. 12-1).

Wide exposures are practical when shingles are laid in double courses. Two grades of shingles may be used—a less expensive grade for the inner course. The first course of shingles in double coursing is commonly tripled. This is done by laying a double undercourse over which the first exterior course is applied. Equal quantities of each kind of shingle are required for the other courses.

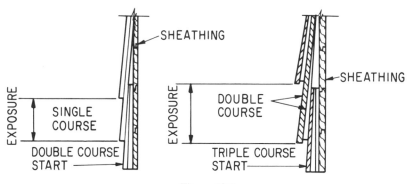

Figure 12-1

THE TAKE-OFF

To determine the quantity of shingles required for a wall, find the total area of the wall including all openings. If the combined area of the openings in a wall is more than 5% of the total area, subtract the combined area of the openings from the total area.

If the combined area of openings is less than 5% of the total wall area (including openings), no deduction is necessary.

In any case, add 5% to the total area for waste in cutting. If deductions are made from the total area, add 5% for waste in cutting.

For example, figure the area of the front wall shown in Fig. 12-2. Windows are 3'0" × 4'6"; the door is 3'0" × 7'0"; and the total area = 36 ft × 8.5 ft = 306 sq ft.

Windows and door area =
$$2 \text{ at } 3 \text{ ft} \times 4.5 \text{ ft} = 27 \text{ sq ft (windows)}$$
$$\underline{3 \text{ ft} \times 7 \text{ ft} = 21 \text{ sq ft (door)}}$$
$$48 \text{ sq ft combined area of openings}$$

5% of total area = 0.05 × 306 sq ft = 15.3 sq ft

It can be seen that the combined area of the openings is more than 5% of the total area. Therefore, subtract the combined area of the openings from the total area, such as

$$\begin{array}{r} 306 \text{ sq ft} \\ - \ \ 48 \text{ sq ft} \\ \hline 258 \text{ sq ft} \end{array}$$

Add 5% for waste in cutting:

258 sq ft × 1.05 = 271 sq ft to be shingled

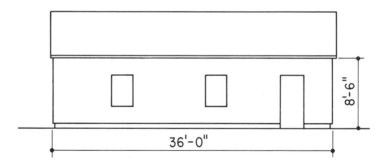

Figure 12-2

If the shingles are to be laid in single courses, divide the square feet to be shingled by the area that can be covered by one bundle. Figure 12-3 gives the area in square feet covered by one bundle of shingles of a particular length and exposure width.

Assume a shingle exposure of 5 in. and a length of 16 in. This size shingle in one bundle will cover 25 sq ft (see Fig. 12-3). Therefore,

271 sq ft ÷ 25 sq ft = 10.8 bundles

To find the additional number of shingles for the double course at the bottom, divide the length of the double course by the length of a single course that can be

Width of Exposure	Length of Shingles		
	16 Inches	18 Inches	24 Inches
4	20	$17\frac{1}{2}$	
$4\frac{1}{2}$	$22\frac{1}{2}$	20	
5	25	$22\frac{1}{2}$	
$5\frac{1}{2}$	$27\frac{1}{2}$	25	
6	30	27	20
$6\frac{1}{2}$	$32\frac{1}{2}$	29	$22\frac{1}{2}$
7	35	31	24
$7\frac{1}{2}$	$37\frac{1}{2}$	34	25
8	40	36	$26\frac{1}{2}$
$8\frac{1}{2}$	$42\frac{1}{2}$	38	28
9	45	40	30
$9\frac{1}{2}$	$47\frac{1}{2}$	43	$31\frac{1}{2}$
10	50	45	33
$10\frac{1}{2}$	$52\frac{1}{2}$	47	35
11	55	50	$36\frac{1}{2}$
$11\frac{1}{2}$	$57\frac{1}{2}$	52	38
12	60	54	40
$12\frac{1}{2}$		57	$41\frac{1}{2}$
13		59	43
$13\frac{1}{2}$		61	45
14		63	$46\frac{1}{2}$
$14\frac{1}{2}$			48
15			50
$15\frac{1}{2}$			$51\frac{1}{2}$
16			53

AREA IN SQUARE FEET COVERED BY ONE BUNDLE OF WOOD SHINGLES

Figure 12-3

laid with one bundle of shingles. Figure 12-4 gives the length of a single course that can be laid by one bundle of shingles.

The length of the double course at the bottom of our building in Fig. 12-2 is 36'0". One bundle of shingles of a four-bundle square, with a 5-in. width of exposure, covers a linear length of 60 ft (see Fig. 12-4).

LENGTH IN FEET OF A SINGLE COURSE WHICH CAN BE LAID BY ONE BUNDLE OF SHINGLES					
Width of Exposure	One Bundle Squares	Two Bundle Squares	Three Bundle Squares	Four Bundle Squares	Five Bundle Squares
4	300	150	100	75	60
$4\frac{1}{2}$	266	133	88	66	53
5	240	120	80	60	48
$5\frac{1}{2}$	218	109	72	54	43
6	200	100	66	50	40
$6\frac{1}{2}$	184	92	63	46	37
7	170	85	57	42	34
$7\frac{1}{2}$	160	80	53	40	32
8	150	75	50	37	30

Figure 12-4

Therefore,

$$36 \text{ ft} \div 60 \text{ ft} = 0.6 \text{ bundle}$$

We have seen that our wall in Fig. 12-2 requires 10.8 bundles. Add the 0.6 bundle for the total bundles required, such as

$$\begin{array}{r} 10.8 \\ +\ \underline{0.6} \\ 11.4, \text{ or } 12, \text{ bundles (total)} \end{array}$$

A word of explanation should be made here in reference to the table in Fig. 12-4: The headings in the table marked "One Bundle Squares," "Two Bundle Squares," and so on, suggests that a one-bundle square will cover one square, which is equal to an area of 100 sq ft. A two-bundle square means that there are two smaller bundles that will cover one square or 100 sq ft. It becomes quite obvious that one bundle of shingles of a four-bundle square contains one-fourth of a one-bundle square and covers only 25 sq ft of surface.

Problem

Find the bundles of wood shingles required on the side walls of the house shown in Fig. 12-5. Doors are 3'0" × 7'0", seven windows are 2'8" × 4'6", and one window is 2'4" × 3'30". Shingles are laid 6 in. to the weather and are 24 in. long.

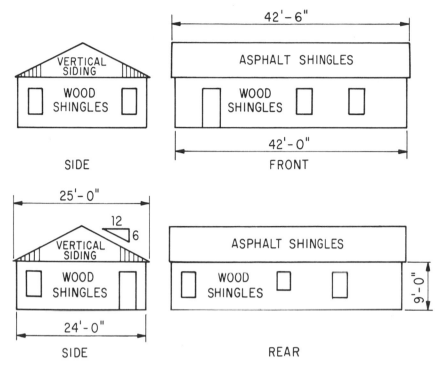

Figure 12-5

Solution Total wall area:

$$42 \text{ ft} + 42 \text{ ft} + 24 \text{ ft} + 24 \text{ ft} = 132 \text{ ft} \times 9 \text{ ft} = 1188 \text{ sq ft}$$

Window and door areas:

Two doors: $3 \text{ ft} \times 7 \text{ ft} = 21 \text{ sq ft} \times 2 = 42 \text{ sq ft}$

Seven windows: $2.67 \text{ ft} \times 4.5 \text{ ft} = 12 \text{ sq ft} \times 7 = 84 \text{ sq ft}$

One window: $2.33 \text{ ft} \times 3 \text{ ft} = 6.99\text{: sq ft} \times 1 = 6.99 \text{ sq ft}$

Combined area of openings $= 132.99$ or 133 sq ft

Total area $= 1188 \text{ sq ft}$

5% of 1188 is $1188 \times 0.05 = 59.4 \text{ sq ft}$
(less than total area of doors and windows)

Since the area of the doors and windows is greater than 5% of the total area, subtract:

$$\begin{array}{r} 1188 \text{ sq ft} \\ - 133 \text{ sq ft} \\ \hline 1055 \text{ sq ft} \end{array}$$

Add 5% for waste in cutting:

$$1055 \text{ sq ft} \times 1.05 = 1107.75, \text{ or } 1108, \text{ sq ft}$$

One bundle of a 24-in. shingle with a 6-in. exposure will cover 20 sq ft (see Fig. 12-3). Therefore,

$$1108 \text{ sq ft} \div 20 \text{ sq ft} = 55.4 \text{ bundles}$$

The addition for a starter course is

$$132 \text{ ft} \div 50 \text{ ft} \text{ (see Fig. 12-4)} = 2.6 \text{ bundles}$$

$$
\begin{array}{l}
55.4 \text{ bundles for walls} \\
\underline{\ 2.6 \text{ bundles for starting course}} \\
58.0 \text{ total bundles required}
\end{array}
$$

Asphalt Shingles

Most asphalt shingles are furnished in two, three, or four bundles to the square. One bundle of a two-bundle square will cover 50 sq ft; one bundle of a three-bundle square will cover $33\frac{1}{3}$ sq ft; and one bundle of a four-bundle square will cover 25 sq ft.

When the area of the roof is determined, add 5% to this area for waste in cutting and fitting the shingles. Then divide the total area by 100 to arrive at the number of squares. The number of bundles of shingles can be easily found by multiplying the number of bundles required to cover one square by the number of squares found.

Problem

A roof measures 42 ft by 22 ft. Find the number of bundles of asphalt shingles when ordered in two-bundle squares. The shingle exposure is $4\frac{1}{2}$ in.

Solution The area of the roof is $42 \times 22 = 924$ sq ft. Add 5% for waste: 1.05×924 sq ft $= 970$ sq ft.

$$970 \text{ sq ft} \div 100 \text{ sq ft} = 9.7 \text{ squares}$$

One bundle of a two-bundle square covers 50 sq ft. Therefore,

$$9.7 \times 2 \text{ (two-bundle square)} = 19.4 \text{ bundles}$$

To find the extra number of shingles required for the starter course, get the length of the starter course and divide by the length of a single course that can be laid by one bundle of shingles. Therefore, one bundle of shingles with a $4\frac{1}{2}$-in. exposure and ordered in two-bundle squares will cover a single course length of 133 ft (see Fig. 12-4). The length of the starter course is 42 ft. Then

$$42 \text{ ft} \div 133 \text{ ft} = 0.32 \text{ bundle}$$

Therefore,

$$\text{Roof bundles} = 19.4$$
$$\text{Starting course bundles} = \underline{0.32}$$
$$19.72, \text{ or } 20, \text{ bundles}$$

Note: Any fractional part of a bundle is counted as one full bundle.

Determining Gable Roof Areas

Roof areas can quickly be found by multiplying the horizontal distance between the lower ends of the gable by the length of the roof (Fig. 12-6). This will yield the area of the horizontal flat. By multiplying the flat area by the percentage factor in the table (Fig. 12-7) for the various roof pitches, the surface area of the roof is found. In Fig. 12-6, on the elevation, the rise of the roof is 9 and the run is 12. In the table in Fig. 12-7 the pitch of a 9–12 roof has a percentage factor of 1.25.

For example, multiply 24 ft × 40 ft = 960 sq ft (flat):

960 sq ft × 1.25 (factor) = 1200 sq ft.

Another and perhaps even simpler method to find the roof area is to scale from the drawings the distance AB shown on the elevation in Fig. 12-6. This distance multiplied by the length BC will yield the area of one side of the roof. Multiply this area by two for the total roof area.

Problem

Find the bundles of asphalt roof shingles for the house shown in Fig. 12-5. Shingles are to be purchased in three-bundle squares, having a $4\frac{1}{2}$-in. exposure.

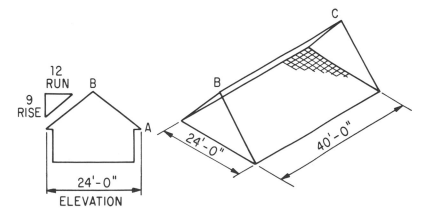

Figure 12-6

Solution   The area of roof is

$$42.5 \text{ ft} \times 25 \text{ ft} = 1062.5 \text{ sq ft (flat, see Fig. 12-6)}$$

$$1062.5 \text{ sq ft} \times 1.118 \text{ (factor)} = 1188 \text{ sq ft}$$

$$\text{Add 5\% } 1188 \text{ sq ft} \times 1.05 = 1247 \text{ sq ft}$$

$$1247 \text{ sq ft} \div 100 \text{ sq ft} = 12.47 \text{ squares}$$

$$12.47 \times 3 \text{ (three-bundle square)} = 37.4, \text{ or } 38, \text{ bundles}$$

The length of the starter course is 42.5 ft + 42.5 ft = 85 ft.   Then 85 ft ÷ 88 ft (see Fig. 12-4) = 0.96 bundle.   Therefore,

$$\text{Roof bundles} = 38$$
$$\text{Starting course bundles} = \underline{\;0.96\;}$$
$$38.96, \text{ or } 39, \text{ bundles}$$

PERCENTAGE FACTOR FOR ROOF AREAS								
Rise	Run	Pitch	Percentage Factor	Rise	Run	Pitch	Percentage Factor
2	12	$\frac{1}{12}$	1.014	16	12	$\frac{2}{3}$	1.667
3	12	$\frac{1}{8}$	1.032	17	12	$\frac{17}{24}$	1.734
4	12	$\frac{1}{6}$	1.054	18	12	$\frac{3}{4}$	1.803
5	12	$\frac{5}{24}$	1.082	19	12	$\frac{19}{24}$	1.873
6	12	$\frac{1}{4}$	1.118	20	12	$\frac{5}{6}$	1 942
7	12	$\frac{7}{24}$	1.158	21	12	$\frac{7}{8}$	2.016
8	12	$\frac{1}{3}$	1.202	22	12	11.12	2.088
9	12	$\frac{3}{8}$	1.25	23	12	$\frac{23}{24}$	2.162
10	12	$\frac{5}{12}$	1.302	24	12	1	2.236
11	12	$\frac{11}{24}$	1.357	25	12	$\frac{25}{24}$	2 311
12	12	$\frac{1}{2}$	1 414	26	12	$\frac{13}{12}$	2.387
13	12	$\frac{13}{24}$	1.474	27	12	$\frac{9}{8}$	2.463
14	12	7.12	1.537	28	12	$\frac{7}{6}$	2.538
15	12	$\frac{5}{8}$	1.601	29	12	29.24	2.615

Figure 12-7

SELF EXAMINATION

Now you have the opportunity to find out how well you understand this unit. Work out the following problem and check your answers with those given in the Answer Box.

For the house shown in Fig. 12-8, find the following: (1) the total area of wall surface to be shingled, including 5% for waste; and (2) the number of bundles of wood shingles for walls, including doubling of first course (use an 8-in. width of exposure and an 18-in. shingle length).

Window and door openings:

Nine windows at 4'0" × 2'8"
One window at 2'8" × 2'0"
One window at 4'0" × 7'0"
Two doors at 3'0" × 7'0"

Answer Box	
Area of wall surface plus 5%	691 sq ft
Bundles of wood shingles for walls and doubling of first course	22 Bundles (four bundles per square)

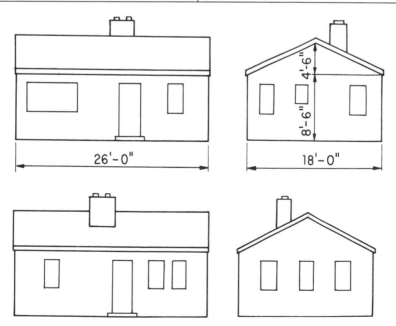

Figure 12-8

ASSIGNMENT—12

1. Find the total number of bundles of wood shingles required for the house shown in Fig. 12-9.   Use a 7-in. width of shingle exposure and a 24-in. shingle length.

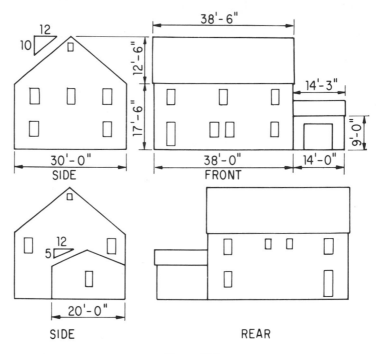

Figure 12-9

2. Find the total bundles of asphalt shingles required for the roof, using three bundles to the square.
3. Find the cost of the material and the installation costs for the shingles found in Problem 2.   Also find the cost with 20% overhead and profit.

17 windows—2'8" × 4'6"

2 windows—1'6" × 2'4"

2 windows—1'8" × 3'4"

2 doors—3'0" × 7'0"

Garage door—8'0" × 7'0"

SUPPLEMENTARY INFORMATION

Data on Asphalt Shingles (Packed two bundles per square in the South.)

Product	Weight per Square (lb.)	Head Lap (in.)	Exposure (in.)	Shingles per Square	Bundles per Square
TAPERED STRIP SHINGLES	275	2	5	80	3
GIANT OVERLAY STRIP SHINGLES	250	2	5	80	3
12" THIKBUT SHINGLES	210	2	5	80	3
12" CEDARTEX THIKBUT SHINGLES	210	2	5	80	3
SUPER GIANT INDIVIDUAL SHINGLES	325	6	5	226	4
DUTCH LAP GIANT INDIVIDUAL	162	2	10×13	113	2
$11\frac{1}{3}$" HEXAGON SHINGLES	167	2	$4\frac{2}{3}$	86	2

Asbestos Shingles

Asbestos shingles are sold by the square, covering 100 sq ft of surface. These shingles are rigid and are made of asbestos fiber and portland cement. They are furnished in numerous styles, sizes, and colors. In estimating the quantities required for a roof, determine the total square feet of roof area and divide by 100 to get the number of squares required.

Lengths of eaves must be measured to obtain the number of linear feet of starters required. Also, the length of hips, valley, and ridges must be found as either shingles or special ridge roll will be required.

Data on Asbestos–Cement Shingles

Product	Weight per Square (lb.)	Head Lap (in.)	Side Lap (in.)	Exposure (in.)	Shingles per Square
 HEXAGONAL SMOOTH FINISH	265	—	3	13×13	86
 DUTCH LAP WOODGRAIN FINISH	280	3	4	12×13	92

Data on Asbestos–Cement Siding

Product	Weight per Square (in.)	Head Lap (in.)	Side Lap (in.)	Exposure (in.)	Shingles per Square
 TAPERTEX STRAIGHT EDGE	185	$1\frac{1}{2}$	—	$10\frac{1}{2} \times 24$	57
 WAVELINE WOODGRAIN FINISH	185	$1\frac{1}{2}$	—	$10\frac{1}{2} \times 24$	57
 STRAIGHT EDGE WOODGRAIN FINISH	185	$1\frac{1}{2}$	—	$10\frac{1}{2} \times 24$	57

    For shed or gable roofs that do not contain hips or valleys, add 5% to the area to allow for waste. Add 2% to the 5% if the roof has hips or valleys.

MATERIALS AND INSTALLATION COSTS

Wood shingles: 16″

> $ 97 per square (100 sq. ft.) for material
>   62 per square for installation
> $149 per square

Clay tile shingles: $\frac{1}{4}$ in. × 11 in lanai

> $270 per square for material
>   88 per square for installation
> $358 per square

Aluminum shingles: .020 in. thick

> $110 per square for material
>   62 per square for installation
> $172 per square

Asbestos shingles: with felt underlay 240 lb./sq. 3 bundle sq.

> $45.35 per square for material
> 32.00 per square for installation
> $77.35 per square

REVIEW QUESTIONS

1. A gable roof with an area of 900 sq ft is to be covered with Lanai clay tile. What is the total cost of the job? Include 28% for overhead and profit.
2. A 900-sq ft roof area is to be covered with 16-in. red cedar wood shingles with a 5-in. exposure. What is the material cost? The installation cost? The total cost with overhead and profit?
3. How are asphalt shingles furnished?
4. What is the weight per square of asphalt tapered strip shingles?
5. One bundle of a two-bundle square covers how many square feet of roof surface?

13

INSULATION: SHEET METAL WORK

INTRODUCTION

This unit deals with estimating the quantity of various types of insulation used in walls between the studding and in ceilings of the house and with the time required to install such insulation. Quantities and labor time of sheet metal for roof flashing, gutters, hangers, conductor pipe, and duct work is included.

Insulation

There are a number of different types of insulation in common use today, particularly the following (Fig. 13-1):

1. Blanket-type insulation consists of fluffed mineral wool placed between two layers of waterproof paper. It is manufactured in widths so that it can be placed between regularly spaced studding or joists without cutting along the sides. These strips of insulation are stapled to adjacent studding or joists through flanges along the edges. It is delivered in rolls or in folded layers.
2. Batt-type insulation consists of a thick pad of fluffed mineral wool attached to a waterproof backing. These batts are also manufactured in widths so that they fit between stud and joist spaces.
3. Granulated- or loose-type insulation is a fluffed material, either mineral or glass wool, that can be poured or blown into walls or placed over plastered

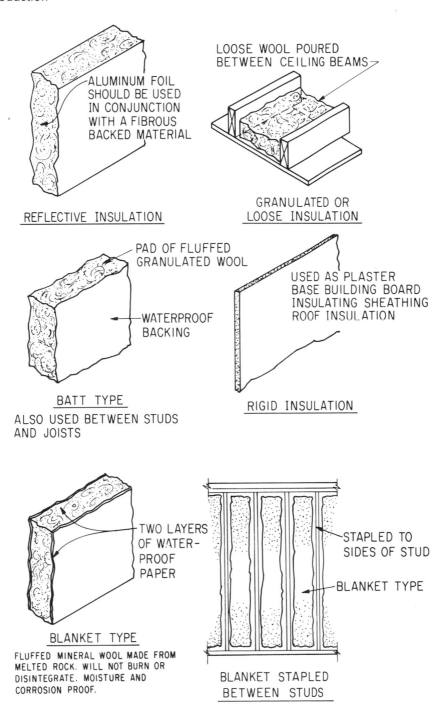

ALUMINUM FOIL
SHOULD BE USED
IN CONJUNCTION
WITH A FIBROUS
BACKED MATERIAL

REFLECTIVE INSULATION

LOOSE WOOL POURED
BETWEEN CEILING BEAMS

GRANULATED OR
LOOSE INSULATION

PAD OF FLUFFED
GRANULATED WOOL

WATERPROOF
BACKING

BATT TYPE
ALSO USED BETWEEN STUDS
AND JOISTS

USED AS PLASTER
BASE BUILDING BOARD
INSULATING SHEATHING
ROOF INSULATION

RIGID INSULATION

TWO LAYERS
OF WATER-
PROOF
PAPER

BLANKET TYPE
FLUFFED MINERAL WOOL MADE FROM
MELTED ROCK. WILL NOT BURN OR
DISINTEGRATE. MOISTURE AND
CORROSION PROOF.

STAPLED TO
SIDES OF STUD

BLANKET TYPE

BLANKET STAPLED
BETWEEN STUDS

Figure 13-1

ceilings between the ceiling joists.   This type of insulation is used in insulating older buildings.   It is packed into bags for delivery.

4. Rigid-type insulation serves as a plaster base, building board, insulating sheathing, and roof insulation.   This insulation is commonly manufactured in sheets 18 in. wide and 32 or 48 in. long.   It is packed in cartons for delivery.

How Much Insulation Is Enough?

The answer depends on the weather where you live.   In Fig. 13-2 is shown a winter heating zone map on which you can locate the zone in which you live and find the *R*-value in Fig. 13-3 for attic, wall, and floor insulation. The *R*-value is the thermal resistance of a building material to the flow of heat.

THE TAKE-OFF

Estimating Quantities of Insulation

Insulation may be ordered by giving the number of square feet needed or by giving the number of packaged units in which the insulation is delivered.   The number of square feet or the number of packaged units of insulation required for a building can be found by calculating the distance around the building, using the exterior dimensions, and multiplying by the ceiling height for a one-story building.   For a

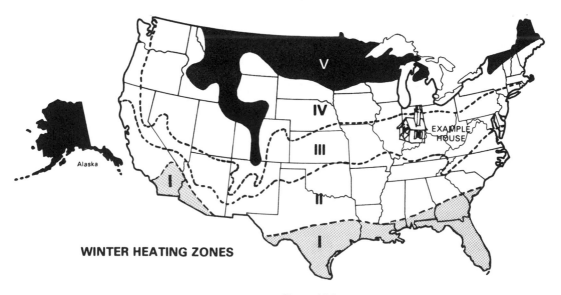

Figure 13-2

R- AND U-VALUES			
Heating Zone	Recommended for:		
	Attic	Wall	Floor
I	R26 U=0.038	R13 U=0.076	R11 U=0.090
II	R26 U=0.038	R19 U=0.052	R13 U=0.076
III	R30 U=0.033	R19 U=0.052	R19 U=0.052
IV	R33 U=0.030	R19 U=0.052	R22 U=0.045
V	R38 U=0.026	R19 U=0.052	R22 U=0.045

Figure 13-3

two-story building, the height from the finished first floor to the second floor ceiling is used.

Quantities of insulation are indicated by the number of square feet of wall surface to be insulated. Information on the area that can be covered by a roll, carton, bag, or other packaged unit can be obtained from the manufacturer or from retail outlets. With this information, the number of rolls, cartons, or bags needed to cover a given area can be easily determined.

If the combined area of openings such as doors and windows is greater than 5% of the total area, subtract the combined area of openings from the total area. Add 8% to the resulting area for waste.

Ceilings

Find the area of the ceiling upon which insulation is to be placed. Add 8% to this area to allow for waste. This is the estimated amount of insulation needed for the ceiling. If one kind of insulation is used on walls and ceilings, divide this combined area by the area of insulation that can be covered by one packaged unit, counting any fractional part of a unit as an additional package.

If one kind of insulation is used on the walls and another type on the ceilings, find the number of packaged units separately from the quantities as found. The table in Fig. 13-4 gives some of the sizes and covering capacities of various types of insulation.

Problem

Find the quantity of roll-blanket insulation for the outside walls of the house shown in Fig. 13-5. The blanket is to be 3 in. thick and 15 in. wide. Also, find the number of bags of pouring wool for all ceilings.

Solution Distance around building:

$$46 \text{ ft} + 38 \text{ ft} + 46 \text{ ft} + 38 \text{ ft} = 168 \text{ ft}$$

$$168 \text{ ft} \times 8 \text{ ft} = 1344 \text{ sq ft}$$

PACKAGED INSULATION, SIZES, AND COVERING CAPACITIES					
Type	Thickness	Width	Length	Pcs. per Pkg.	Sq. ft. per Pkg.
Roll Blanket	$1\frac{1}{2}$"	*48"	50'	1	200
		ᵀ15"	80'	1	100
	2"	15"	60'	1	75
		23"	52'	1	100
	3"	15"	72"	1	90
Batt Blanket	$1\frac{1}{2}$"	15"	96"	8	80
	2"	15"	24"	24	60
		23"	48"	8	$60\frac{1}{2}$
	3"	15"	24"	16	40
		15"	48"	8	40
Pouring Wool	Packaged in bags. Each bag will cover approximately 25 sq. ft. of a 4" thickness.				

* Available in Western section of U.S.A.
ᵀ Not available in Western section of U.S.A.

Figure 13-4

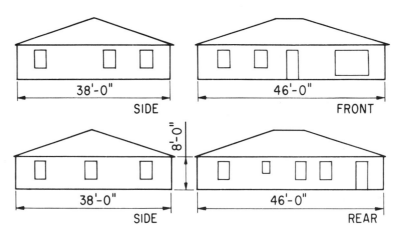

Figure 13-5

Window and door area (combined):

Windows: 9 at 2.5 ft × 4.5 ft × 9 =   100.8 sq ft
          1 at 2 ft × 3 ft × 1 =   6 sq ft
          1 at 6 ft × 8 ft × 1 =   48 sq ft
Doors: 2 at 3 ft × 7 ft × 2 =   42 sq ft
                                        196.8 sq ft
                                  combined openings

5% of 1344 sq ft = 0.05 × 1344 sq ft = 67.2 sq ft

Since the combined window and door area is greater than 5% of the total area, subtract the combined opening area from the total area:

1344.0 sq ft

− 196.8 sq ft

1147.2 sq ft

Add 8% for waste:

1147.2 sq ft × 1.08 = 1239 sq ft net wall area

Refer to Fig. 13-4 for the covering capacities of the insulation. A 3-in. thick roll blanket, 15 in. wide in one package, will cover 90 sq ft; therefore

1239 sq ft ÷ 9 sq ft = 14 packages of insulation

Ceilings: Area is 46 ft × 38 ft = 1748 sq ft

Add 8% for waste: 1748 sq ft × 1.08 = 1888 sq ft

Refer to Fig. 13-4, which gives the covering capacity of pouring wool that is 25 sq ft per bag of 4-in. thickness; then

1888 sq ft ÷ 25 sq ft = 76 bags for ceilings

Sheet Metal Work

Sheet metal is figured by the weight of material used and the time required to install the metal. There are various gauges of sheet metal ranging from $\frac{1}{2}$ to $\frac{1}{160}$ in.; the gauges ranging from $\frac{1}{16}$ to $\frac{1}{64}$ in. are the most common for roofing and flashing.

Figure 13-6 gives the most common U.S. standard sheet metal gauges and the weights per square foot of metal.

Sheet metal for roof ridging. Most estimators or contractors scale the length of all roof ridges from the drawings and in that manner find the linear feet required. Sheet metal for ridging is commonly 13 in. wide and comes in 8-ft lengths. Some overlap should be considered.

UNITED STATES STANDARD SHEET METAL GAUGES		
Number of Gauge	Thickness in Decimals of an Inch	Weight per sq. ft. in lb.
16	0.0625	2.5
17	0.0562	2.25
18	0.05	2
19	0.0437	1.75
20	0.0375	1.5
21	0.0343	1.375
22	0.0312	1.25
23	0.0281	1.125
24	0.025	1.000
25	0.0218	0.875
26	0.0187	0.75
27	0.0171	0.6875
28	0.0156	0.625

Figure 13-6

Problem

Assume that the scaled lengths of all roof ridges for a house add up to 108 lin ft. For 100 ft of length allow about 3 ft additional for overlap and waste. The total length, therefore, is 111 ft. Assuming that a 26-gauge metal is used, find the total weight and the time required to install this metal.

Solution   111 ft is the total length. 26-gauge metal from Fig. 13-6 indicates a weight of 0.75 lb/sq ft. To find the total square feet, multiply the length by the width:

$$111 \text{ ft} \times 108 \text{ ft (13 in.)} = 120 \text{ sq ft}$$

Then

$$120 \text{ sq ft} \times 0.75 \text{ lb (weight per sq ft)} = 90 \text{ lb}$$

A sheet metal worker can install 20 lin ft of roof ridging per hour; therefore

$$111 \text{ ft} \div 20 \text{ ft} = 5\tfrac{1}{2} \text{ hr to do the job}$$

Sheet metal for roof valleys.   All valley lengths are scaled from the drawings. Some allowance for overlapping the metal should be made, generally 2 in. for each 8 ft of length. Valley flashing can be installed at the rate of 27 ft/hr. It should be quite evident that valley lines on the plans are not shown in true length and cannot be scaled from the drawings as such. However, the valley line true lengths can be found readily by a few simple manipulations. Look at Fig. 13-7— a plan view of two intersecting roofs and an elevation view. The valley line AB is not in true length on the plan. Line CD in the elevation is the true length of line CD on the plan. Line CD on the elevation is not the same length as line AB on the plan. Line AB actually is longer.

To find the true length of line AB, place the needle point of a compass on point A and the pencil point of the compass on point B and scribe the arc BF. Project point F straight down to the roof edge of the elevation to point P. Connect point A with point P, which becomes the true length of the line AB. Line DP then can be scaled off as the correct length of line AB.

Estimating the Amount of Gutter Material

All lengths of gutter are scaled from the blueprints to determine the linear feet required. The time required to install gutters can vary widely, depending on the lengths of each gutter run. Where a great many short gutter runs are put up the

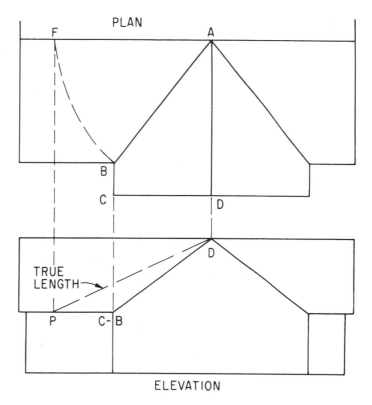

Figure 13-7

labor time might be as much as 1 hr/15 lin ft of gutter. This is due to the fact that the ends of gutters have to be closed; on short runs this requires considerably more time than on long runs. Where gutter runs are long, one worker can put up as much as 40 lin ft/hr. On the average residence, 17 ft of gutter can be installed by one worker in 1 hr.

Gutter hangers. Gutters are supported in intervals of 3 ft. To find the number required, divide the total length of gutter by the spacing of hangers (3′0″) and add four to six for losses. The labor to install hangers is included in the gutter time.

Conductor pipe. This is figured by scaling the length of pipe from the elevation wherever such pipe is indicated or where specifications call for conductor pipe. Just as in gutters, the labor time for conductor pipe varies. A long run requires less time per linear foot than a short run. A conductor pipe must be soldered to the gutter and often fitted into a sewer pipe connection. For the average residence, about 10 lin ft of conductor pipe can be installed in 1 hr.

MATERIAL AND INSTALLATION COSTS

Note: Percentages for overhead and profit are assumed; they may vary with each subcontractor and locality.

Ductwork: rectangular, including joints and fittings and supports, but without insulation.

$1.20 per square foot of metal surface, for material

<u>2.04</u> per square foot of metal surface, for installation
$3.24 per square foot of metal surface installed

Assuming 32% for overhead and profit = $4.27 per square foot of metal surfaces installed.

Conductor pipe (downspouts): regular, 3 in. × 4 in. corrugated stock. $9.50 per linear foot, installed. Assuming 14% for overhead and profit = $10.83 per linear foot.

Gutters: Galvanized steel, half-round 28 gauge and 5 in. wide costs $2.15 per linear foot in place. Assuming 14% for overhead and profit = $2.45 per linear foot.

Flashing: aluminum 0.013 thick $1.50 per square foot
                           0.032 thick $2.25 per square foot

Insulation: Poured fiberglass wool *R*4 per inch costs $1.5 per cubic foot. Mineral wool *R*3 per inch costs $1.43 per cubic foot in place.

Fiberglass batts or blankets: $3\frac{1}{4}$-in.-thick $R11$, 15 in. wide, cost $0.25 per square foot in place.

SELF EXAMINATION

Now you have the opportunity to find out how well you understand the material in this unit. Work out the following problem, and check your answers with those given in the Answer Box.

Estimate the quantity of material and the time required for installation of the following (refer to Fig. 13-8):

1. Sheet metal for roof ridge and valleys 13 in. wide, 24 gauge.
2. Gutters, hangers, and downspouts and their costs in place.

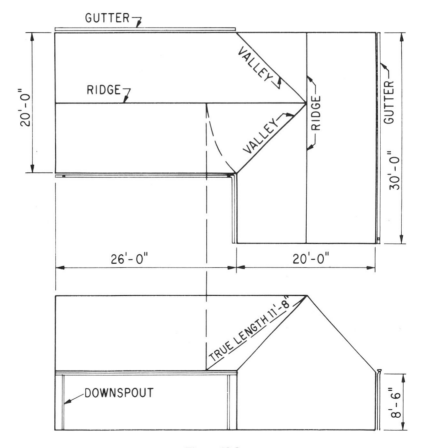

Figure 13-8

Answer Box		
Item	Quantity	Labor Time (hr)
Ridge	73.44 lb	3.3
Valleys	25.27 lb	1.3
Gutters	92 lin ft	5.4
Hangers	30.7 or 31 Add 4 = 35	
Downspouts with elbows and shoes	34 lin ft	3.4

ASSIGNMENT—13

1. A small bungalow-type dwelling, 34 ft long by 28 ft wide, is to be insulated with roll blanket-type insulation between the studs of the outside walls. The floor to ceiling height is 8'0". The roll blanket is 2 in. thick and 15 in. wide. In this problem, consider the openings take care of the allowances for waste. Find the amount of insulation required.
2. Figure the bags of pouring wool for the ceiling of the bungalow. Add 8% for waste.
3. What is the R-value of a 1.13 U-factor?
4. What is the cost of a cubic foot of poured fiberglass wool R4 per inch of insulation?

REVIEW QUESTIONS

1. Name four types of insulation.
2. Explain how quantities of insulation are estimated.
3. A 2-in.-thick roll blanket 60 ft long contains how many square feet per package?
4. A 16-gauge piece of sheet metal weighs how many pounds?
5. How is the amount of gutter material estimated?

14

PAINTING

INTRODUCTION

In this unit you will learn how to estimate the quantities of paint required for various surfaces and the usual time it requires to apply the paint to these surfaces. Often, instead of painting interior walls, contractors will apply wallpaper, at the request of the client. Estimating rolls of wallpaper therefore also forms a part of this unit.

Paint is used primarily as a protective coating against all kinds of weathering conditions, over wood, metal, concrete, and numerous other materials. In addition, paint enhances the appearance of a surface through the introduction of oxide (colors) into the paint.

Almost all paint today is a factory-mixed product that is tinted by the retailer to the color desired.

The best method of estimating paint quantities is to find from the plans the actual surface area to be painted or to take actual field measurements. Painting costs are based on two major items: labor and materials. The cost of material depends on the grade or quality of the paint and the quantity used, whereas labor depends on the present-day labor scales.

THE TAKE-OFF

In estimating the amount of paint required for clapboards or wood siding, the actual area of the exterior surface is calculated by finding the combined area of all walls or surfaces.   Each side of the house must be included.

For a square or rectangular house the distance around the house can be multiplied by the height measured from the foundation to the eaves of the roof. Do not deduct for window areas less than 10 ft × 10 ft.   Add 10% to the surface area found.

One gallon of flat acrylic prime coat paint will cover 275 sq ft when applied with a brush.   When this same prime coat is applied with a roller, only 250 sq ft can be covered.   Using a spray gun, however, 325 sq ft can be covered.

If a second coat is to be applied, add 25 sq ft to the areas covered by brush and roller, but no additional coverage is gained by spraying.

In one hour, an experienced painter can apply:

144 sq ft by hand

175 sq ft by roller

500 sq ft by spraying

Problem

How many square feet of flat acrylic prime coat paint can be applied by brush on the house shown in Fig. 14-1, and how many gallons of paint and hours are required to do the job?

Solution   The total number of square feet is found as follows:

$$40 \text{ ft} + 20 \text{ ft} + 40 \text{ ft} + 20 \text{ ft} = 120 \text{ ft}$$

$$120 \text{ ft} \times 9 \text{ ft (height of siding)} = 1080 \text{ sq ft}$$

$$1080 \text{ sq ft} \times 1.10 \text{ (plus 10\%)} = 1188 \text{ sq ft total}$$

The total number of gallons required is

$$1188 \text{ sq ft} \div 275 \text{ (sq ft/gal)} = 4.32 \text{ gal}$$

The total number of hours required is

$$1188 \text{ sq ft} \div 144 \text{ (sq ft/hr)} = 8.25 \text{ hr}$$

The calculations above may be taken-off in tabular form:

CLAPBOARDS OR SIDING (275 SQ FT/GAL OF ACRYLIC PAINT, 144 SQ FT/HR)

Item	Unit	Length	Height	Square feet	Gallons	Hours
Walls	1	120	9	1080		
		Add 10%		108		
				1188	4.32	8.25

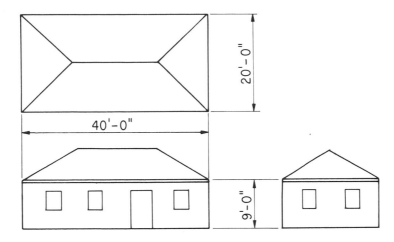

Figure 14-1

Painting the Window Sash and Frames

The window sash and frames are painted at the same time as the walls and the amount of paint figured for the walls is ample for the window sash and frames. No additional paint need be considered.

On brick buildings, however, window sash and frames are figured as separate jobs because the exterior brick walls require no paint. About 3 qt of paint per coat is required for 20 window sash and frames.

Painting Exterior Doors and Door Frames

Doors and frames are painted together with outside wall surfaces, and no additional paint is required. In brick buildings doors and door frames are considered the same as window openings. Therefore, all doors and frames are counted as openings. Paint quantity and time is the same as for windows: 3 qt of paint per coat for 20 doors and frames and 6 hr to do the job.

When window and door frames are painted a contrasting color, no additional paint need be figured because the exterior paint can be tinted in sufficient amount for frame color. 20 openings can be painted in 6 hr. The amount of tinted paint can be figured on the basis of 3 qt for 20 openings.

Problem

For the house shown in Fig. 14-2, figure the following:

1. Total square feet of wall surface.
2. Gallons of acrylic paint by brush required on the wood siding walls.
3. Amount of paint to be tinted for all openings.
4. Number of hours required to do the job.

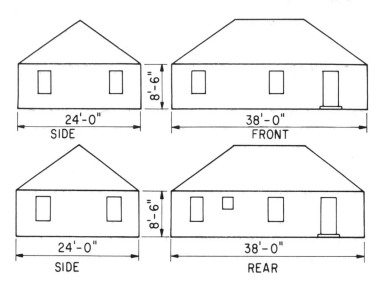

Figure 14-2

Solution The total number of square feet is found as follows:

$$38 \text{ ft} + 24 \text{ ft} + 38 \text{ ft} + 24 \text{ ft} = 124 \text{ ft}$$

$$124 \text{ ft} \times 8.5 \text{ ft (height)} = 1054 \text{ sq ft}$$

$$1054 \text{ sq ft} \times 1.10 \text{ (plus 10\%)} = 1159.4, \text{ or } 1160, \text{ sq ft total wall surface}$$

The total number of gallons is

$$1160 \text{ sq ft} \div 275 \text{ (sq ft/gal)} = 4.21 \text{ gal}$$

The tinted paint amount is as follows. There are 11 window and door openings; since 3 qt are required for 20 openings, 1 opening will require:

$$\tfrac{1}{20} \text{ of 3, or } \tfrac{3}{20}, \text{ qt or } 3 \div 20 = 0.15 \text{ qt/opening}$$

Eleven openings will therefore require

$$11 \times 0.15 \text{ qt} = 1.65, \text{ or } 2, \text{ qt or } \tfrac{1}{2} \text{ gal}$$

The number of hours to do the job is

$$1160 \text{ (total sq ft)} \div 144 \text{ (sq ft/hr)} = 8 \text{ hr}$$

Painting Steel Sash

Steel sash requires about one-third the amount of paint required for wood sash. One quart of paint is ample for one coat on 20 steel sashes.

On new steel sash, one worker can paint six ordinary-sized sashes per hour. On old sash that requires repainting, about 25% more time should be allowed.

Painting Brick and Concrete Walls

In estimating the amount of paint and time required to paint brick or concrete walls, calculate the areas of the surfaces to be covered. Do not deduct for openings unless there are very large openings. One gallon of paint covers about 200 sq ft by brush. One painter can paint about 106 sq ft/hr by brush.

Painting Concrete Floors

In general, paint for concrete floors covers about 400 sq ft/gal for the first coat. One worker can paint about 200 sq ft/hr.

Painting Interior Walls and Ceilings

The areas of the walls are calculated by multiplying the length of each wall by its height. Ceiling areas are found by multiplying the length by the width. Deductions for windows and doors can be made if they are not to be painted. One gallon of flat paint on walls covers 450 sq ft/gal. It takes one worker to paint 200 sq ft/hr.

Painting Interior Doors

Interior paint on doors covers about 450 sq ft/gal for the priming coat and 475 sq ft for second and third coats. The areas of doors are figured by multiplying the length by the width and then multiplying by 2 because of two sides. Some contractors add two or three additional square feet to take care of the bevels on panels.

Painting Striated Shingled Walls

One gallon of latex paint covers about 150 sq ft. It can be applied by one painter in 1 hr.

Varnish or Shellac on Floors

One gallon of varnish covers 400 sq ft of floor area. One painter can apply 300 sq ft in 1 hr.

Glue Size

Dry wall coverings such as sheetrock and wet wall covering such as plaster must receive a sealer coat or "glue size" consisting of flake glue dissolved in water. One gallon of glue size will cover 655 to 700 sq ft and it takes 1 hr to apply.

Setup and Cleanup

The time given in the foregoing is actual painting time. Additional allowance should be made for preparation before starting to paint and for cleaning up after the painting has been completed. Often surfaces must be prepared for painting, such as sanding rough surfaces. This is a factor that must be considered by each contractor individually.

Estimating Wallpaper

One roll of wallpaper contains 36 sq ft. The standard roll measures 18 in. wide and 24 ft long, or 1′6″ wide by 24 ft long = 36 sq ft, which is 4 sq yd. A double roll of wallpaper measures 18 in. wide and 48 ft long, or 1′6″ wide by 48 ft long = 72 sq ft or 8 sq yd. To find the number of rolls of wallpaper required for a room, find the total area of the room by multiplying the length of all the combined walls by the height of the walls. Find the area of all windows and doors, and deduct this area from the wall areas found. The result will be the total number of square feet to be covered. If single rolls are to be used, divide the total number of square feet by 36. If double rolls are to be used, divide the total square feet by 72. It is customary to allow between 20 to 25% for waste and matching.

Problem

A room measures 20 ft × 30 ft and has a ceiling height of 8 ft. There are two doors 2′6″ × 7′0″ and four windows measuring 2′6″ × 4′6″. How many double rolls are required? How many single rolls?

Solution The combined length of all walls is

$$20 \text{ ft } + 30 \text{ ft } + 20 \text{ ft } + 30 \text{ ft } = 100 \text{ ft}$$

$$100 \text{ ft } \times 8 \text{ ft (height of wall)} = 800 \text{ sq ft}$$

Doors:

$$2.5 \text{ ft } \times 7 \text{ ft } = 17.5 \text{ sq ft } \times 2 \text{ (two doors)} = 35 \text{ sq ft}$$

Windows:

$$2.5 \text{ ft } \times 4.5 \text{ ft } = 11.25 \text{ sq ft } \times 4 \text{ (four windows)} = 45 \text{ sq ft}$$

Total combined (windows and doors) = 80 sq ft. Therefore,

$$
\begin{array}{r}
800 \text{ sq ft} \\
- \underline{80 \text{ sq ft}} \\
720 \text{ sq ft}
\end{array}
$$

Allow 15% for waste and matching:

$$720 \text{ sq ft } \times 1.15 \ (15\%) = 828 \text{ sq ft of wallpaper required}$$

$$828 \text{ sq ft } \div 72 \text{ sq ft (sq ft of double roll)} = 11.15 \text{ rolls}$$

$$11.5 \text{ double rolls} = 23 \text{ single rolls}$$

MATERIAL AND INSTALLATION COSTS

> *Walls and ceilings:* painted by roller, costs about $0.18 per square foot. About 1300 sq ft can be covered by one painter in 8 hr.
>
> *Wood siding:* including puttying, where required, costs about $0.36 per square foot. About 670 sq ft can be covered by one painter in 8 hr.
>
> *Brick or concrete:* painted by brush, costs about $0.44 per square foot. About 485 sq ft can be covered by one painter in 8 hr.
>
> *Wood trim:* including puttying, where required, costs about $0.52 per square foot. About 340 sq ft can be covered by one painter in 8 hr.
>
> *Door and frame:* receiving a prime coat plus another coat costs about $18 each. One painter can cover about 11 doors in 8 hr.
>
> *Wallpaper:* for a double roll, the cost is $0.56 per square foot. A paper-hanger can apply about 522 sq ft in 8 hr.

Note: Prices are material and installation averages and should not be taken as actual cost in your area.

SELF EXAMINATION

Find the number of single rolls of wallpaper required for the living room and the two bedrooms shown in Fig. 14-3. Check your answers with those given in the Answer Box. The windows are 2'6" × 4'6". The bedroom doors are 2'6" × 6'8". The closet doors are 2'0" × 6'8". The outside door is 3'0" × 6'8".

Answer Box
Net square feet = 1033 sq ft
Number of single rolls = 29

ASSIGNMENT—14

Estimate the following for flat acrylic paint (Fig. 14-4):

1. Gallons of paint for wood siding.
2. Amount of paint required if all sash and frames, and doors and frames were to be painted only.
3. Quantity of paint required for walls and ceilings in bedrooms 1 and 2 and living room.
4. Quantity of varnish or shellac on floors in living room and bedrooms 1 and 2.

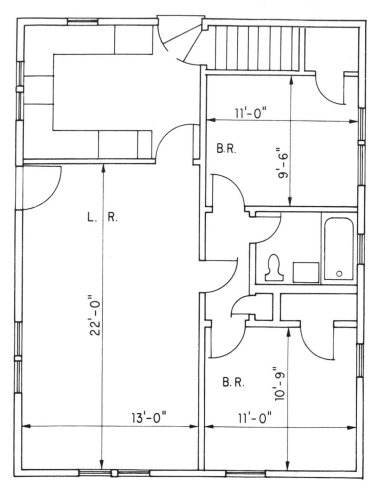

Figure 14-3

SUPPLEMENTARY INFORMATION

SOLVENTS

Type	Type of vehicle
Acetone	Synthetic resins
Alcohol	Shellac, some drying oils, and natural resins
Ethyl alcohol	Drying oils, natural and synthetic resins
Ketone	Drying oils, natural and synthetic resins
Mineral spirits	Drying oils, natural and synthetic resins
Water	Water- and emulsion-base vehicles
Xylene	Water- and emulsion-base vehicles
Turpentine	Drying oils, natural and synthetic resins

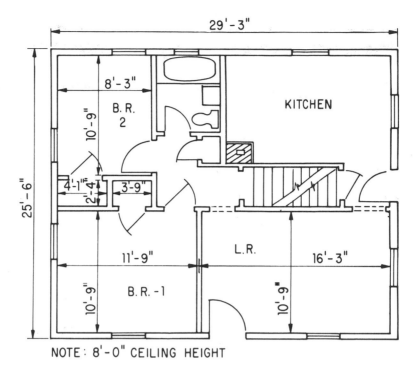

NOTE: 8'-0" CEILING HEIGHT

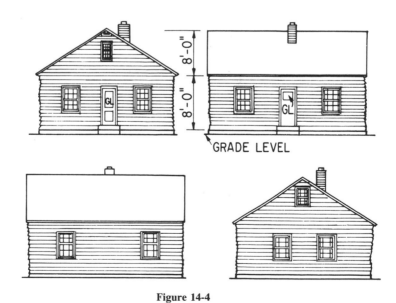

GRADE LEVEL

Figure 14-4

Following are paint prices per gallon for 5-gal lots.   Prices are for comparative purposes only and are approximate and differ widely in different areas.

Acoustical paint	$10.00
Brick, or concrete wall, acrylic	10.50
Creosote	3.31
Epoxy, concrete enamel	20.33
Exterior acrylic latex, one coat	12.57
Shingle paint	8.36
Shingle or wood stain	6.68
Underwater, premium	10.90
Interior latex, base flat, one coat	10.45
Concrete floor paint	10.37
Metal paint, red, quick-drying	19.47
Primer, plaster, wood or metal, oil base	9.09
Swimming pool, rubber base 350 sq ft/gal	21.40
Turpentine	5.35
Varnish, polyurethane	11.52
Exterior spar varnish	12.84
Concrete or masonry, clear	7.11

REVIEW QUESTIONS

1. One gallon of flat acrylic paint applied on wood siding will cover how many square feet?

2. How many square feet of surface can be applied by an experienced painter in Problem 1?

3. How much paint is required to paint 20 doors and frames?   How long does this take?

4. One gallon of latex paint applied on striated shingles will cover how many square feet?

15

DOORS AND WINDOWS

INTRODUCTION

This unit deals with the estimating of doors and windows. Doors are used primarily for exterior entrances and exits from a building and to close off or subdivide areas within a building.

Doors

The materials most generally used in the manufacture of doors are wood, steel, aluminum, stainless steel, glass, fabrics, and plastic. Doors can be identified by their method of opening, such as swinging, folding, sliding, bypassing, overhead, revolving, and rolling (see Fig. 15-1).

The most widely used doors are the wood swing type. They are available in $1\frac{3}{8}$- and $1\frac{3}{4}$-in. thicknesses, $6'8''$ and $7'0''$ in height, and in widths from $1'10''$ to $3'0''$ in 2-in. increments.

The greatest use of steel doors in building construction is for interior doors requiring fire ratings and for entrance and exit doors with or without fire ratings.

Aluminum doors are used for entrance doors, which include the frames, and for interior swing-type doors where no fire ratings are required. They are used for sliding, swing-type, bypassing, revolving, and overhead doors.

Glass doors of tempered glass are used for swing-type doors. This kind of

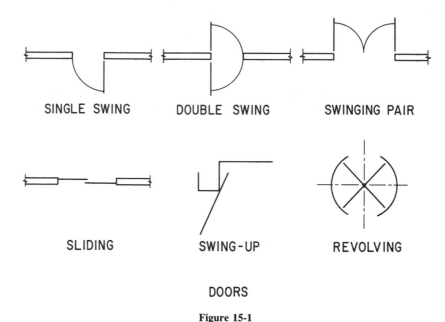

SINGLE SWING   DOUBLE SWING   SWINGING PAIR

SLIDING   SWING-UP   REVOLVING

DOORS

Figure 15-1

glass requires that all pivots, locks, holes and so on, be part of the casting when the tempered glass is manufactured.

To control heat loss or heat gain, many types of storm-door combinations are available. These include single-glazed, double-glazed, and fixed or removable glazed areas. The door frames may be of wood, steel, aluminum, and stainless steel.

Windows

Generally, wood or metal stock windows are standardized as to sizes and are manufactured as complete units that include hardware, weather stripping, operating equipment, and so on. They are furnished either glazed or unglazed and are made to receive window glass, plate glass, special glass, or insulating glass, and they may also be obtained with screens and storm sash. Typical window types are shown in Fig. 15-2.

Windows are manufactured in four categories based on their use in building construction: residential, commercial, industrial, and monumental. All types have established standards that have been developed by window manufacturers' associations and institutes, U.S. Department of Commerce, American National Standards Association, and other city, state, and federal agencies.

All types of windows, when assembled and ready to be delivered to a building

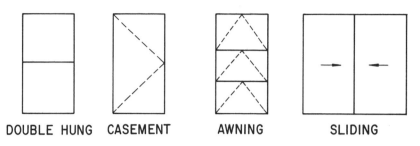

DOUBLE HUNG   CASEMENT   AWNING   SLIDING

TYPICAL WINDOW TYPES

Figure 15-2

under construction or to a warehouse and distributor, receive a factory-finishing treatment, which may be either protective, preservative, permanent finish, or a base for painting, staining, or other type of finish.

Wood window standards have been established for double-hung, casement, sliding, and awning windows. These standards cover materials, sizes, frame, sash, check rails, hardware, operating mechanisms, weatherstripping, screens, and storm sash. The standards require that all wood parts receive preservative treatment against fungi, insects, and water.

Ponderosa pine, kiln-dried to a moisture content of 6 to 12%, is the wood generally used in the manufacture of wood windows. Of all wood window types, the double-hung window is the most widely used.

Steel windows are manufactured as a complete unit including hardware, weather stripping, and operating mechanism, but they are always glazed on the site. Standards have been established regarding sizes, material, weather stripping, hardware, and operating equipment for residential, commercial, industrial, and monumental windows.

Aluminum window extrusions are used to manufacture all the various types and grades of aluminum windows. A minimum of 0.062 in. has been established for material or section thickness. Joints are either welded or fastened by mechanical means. Fasteners and anchors must be of a metal that is compatible or isolated from aluminum so that no *galvanic action** can occur.

Stainless steel window types and grades are produced by a limited number of manufacturers. Stainless steel is available in various finishes, ranging from highly polished to a dull sand-textured finish. It is highly corrosion resistant and exceptionally strong in thin gauges.

---

*When different metals or alloys are in contact with each other and moisture is present, an electric current starts to flow from one metal to the other and in time one of the metals is eaten away while the other remains intact. This process is known as "galvanic action."

DOOR ESTIMATE					
Item	Unit Mat'l. Cost	Unit Install. Cost	Total Mat'l. and Install. Cost	Qty.	Total Cost
① Ext. solid core, birch 3 ft x 7 ft, $1\frac{3}{4}$ in. thick	90	24	114	1	114
② Int. hollow core, birch 3 ft x 7 ft, $1\frac{3}{8}$ in. thick	30	19	49	4	196
③ Sliding patio, aluminum 12 ft x 7 ft, 1 in. tempered glass	695	115	810	1	810
④ Garage door, overhead 16 ft x 7 ft	290	32	322	1	322
			Total cost =		$ 1442
			with 20% O. & P. =		$ 1730

Figure 15-3

THE TAKE-OFF

Estimating the number of doors from a plan is a relatively simple matter, especially for residential constructions.   All that needs to be done is to count the number of exterior and interior doors and to list the quantity by size, material, unit, and installation costs.

Counting the number of doors on plans of multistoried buildings such as large office structures, hospitals, or apartment buildings can become difficult unless, through a systematic colored-pencil check-off on the plans the doors can be spotted and listed.   Plans with a complete door schedule do, of course, simplify the entire take-off.

The exterior and interior door take-off and estimate in Fig. 15-3 is that for the small residential plan (Fig. 15-4).   The unit door cost and the cost of installation is added to arrive at a combined unit and installation cost.   This, multiplied by the quantity or number of doors, will give the total cost.

The subcontractor's overhead and profit was assumed to be approximately an added 20%.   Overhead and profit varies with most contractors based on location and local labor rates.

Check the take-off estimate for exterior and interior doors with the plan to

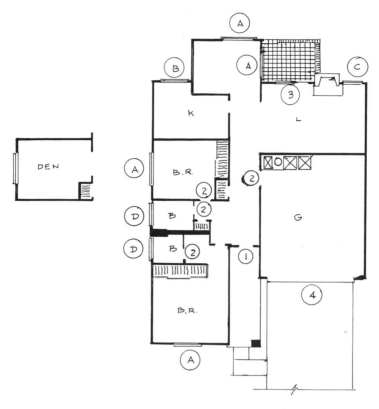

Figure 15-4

see that all doors are itemized.   The unit and installation costs are national average costs.

The take-off for windows is similar to the door take-off.   All windows are itemized by size and type.   The tabular take-off in Fig. 15-5 lists the window corresponding to the letters on the plan.

MATERIAL AND INSTALLATION COSTS

Doors: Exterior solid core doors, 3 ft $\times$ 7 ft $\times$ $1\frac{3}{4}$ in. costs about \$114 for material and installation.   Interior hollow-core doors of birch cost about \$49 for material and installation.   Sliding patio aluminum doors, 12 ft $\times$ 7 ft $\times$ 1 in. with tempered glass, cost about \$810 for the door and its installation.   Garage door, overhead, wood, 16 ft $\times$ 7 ft, costs about \$322 for the door plus its installation.

WINDOW ESTIMATES					
Item	Unit Mat'l. Cost	Unit Install. Cost	Total Mat'l. and Install. Cost	Qty.	Total Cost
Ⓐ Sliding alum., frame, trim, 9 ft x 5 ft opening	173	83	256	4	1024
Ⓑ Sliding alum., frame, trim, 8 ft x 4 ft opening	115	54	169	1	169
Ⓒ Sliding alum., frame, trim, 3'4" x 4'0" opening	80	42	122	1	122
Ⓓ Sliding alum., frame, trim, 5 ft x 3 ft opening	94	36	130	2	260

Total cost = $ 1573
with 20% O. & P. = $ 1888

Figure 15-5

Windows: The costs for a sliding window of aluminum frame and trim and its installation are:

9 ft × 5 ft opening: about $256

8 ft × 4 ft opening: about $182

3′4″ × 4′0″ opening: about $86

5 ft × 8 ft opening: about $227

Note: The prices above do not include overhead and profit.

Wood windows, double hung, 2′8″ × 4′6″, with glass, cost $85 plus $35 for installation. Casement wood windows, including screens, 1′10″ × 3′2″, with glazing, cost $65 plus $35 for installation.

SELF EXAMINATION

From the plan shown in Fig. 15-6 take-off the doors and windows in tabular form and find the total cost of material and installation, plus overhead and profit. Check your answers with those given in the answer box.

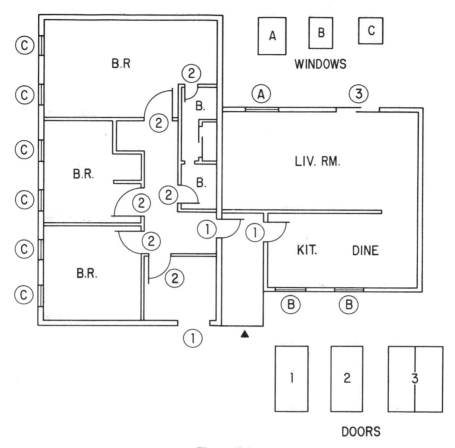

Figure 15-6

Window schedule:

    (A) 9 ft × 5 ft aluminum sliding   1

    (B) 8 ft × 4 ft aluminum sliding   2

    (C) 4′ × 3′4″ aluminum sliding   6

Door schedule:

    (1) 3 ft × 7 ft × $1\frac{3}{4}$ in. exterior   2

    (2) 3 ft × 7 ft × $1\frac{1}{2}$ in. interior   6

    (3) 12 ft × 7 ft sliding aluminum   1

Answer Box	
Windows	$855 with 20% overhead and profit = $1026
Doors	$2256 with 20% overhead and profit = $2707.20

ASSIGNMENT—15

Take-off the windows and doors in tabular form from the plan in Fig. 15-7. Find the total cost of both doors and windows by getting prices of the items from your local lumber dealer.   Refer to Figs. 15-3 and 15-4.

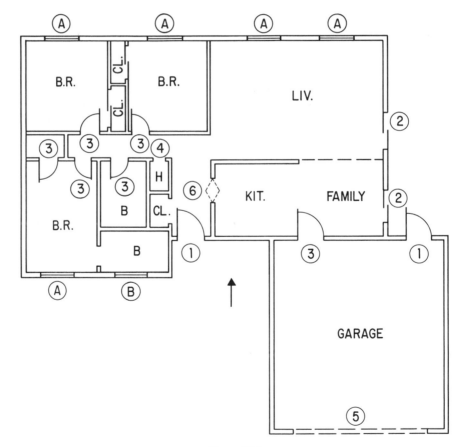

Figure 15-7

Note:
　Windows

　　(A)　Windows are $8'0'' \times 5'0''$

　　(B)　Window is $3'0'' \times 5'0''$

　Doors

　　(1)　$= 3'0'' \times 7'0''$ solid birch

　　(2)　$= 12'0'' \times 6'10\frac{3}{4}''$ aluminum sliding (frame size)

　　(3)　$= 2'8'' \times 7'0'' \times 1\frac{3}{8}''$ hollow core

　　(4)　$= 2'4'' \times 6'8'' \times 1\frac{3}{8}''$ hollow core

　　(5)　$= 16'0'' \times 8'0''$ garage, wood, overhead

　　(6)　$= 2\text{-}2'4'' \times 6'8'' \times 1\frac{3}{8}''$ double acting, hollow core

SUPPLEMENTARY INFORMATION

ALUMINUM SLIDING GLASS DOORS

Glass size (in.)	Frame size	Rough opening
$33 \times 76\frac{3}{4}$	$6'0'' \times 6'10\frac{3}{4}''$	$6'0\frac{1}{2}'' \times 6'11\frac{1}{4}''$
$45 \times 76\frac{3}{4}$	$8'0'' \times 6'10\frac{3}{4}''$	$8'0\frac{1}{2}'' \times 6'11\frac{1}{4}''$
$57 \times 76\frac{3}{4}$	$10'0'' \times 6'10\frac{3}{4}''$	$10'0\frac{1}{2}'' \times 6'11\frac{1}{4}''$
$33 \times 76\frac{3}{4}$	$9'0'' \times 6'10\frac{3}{4}''$	$8'0\frac{1}{2}'' \times 6'11\frac{1}{4}''$
$45 \times 76\frac{3}{4}$	$12'0'' \times 6'10\frac{3}{4}''$	$12'0\frac{1}{2}'' \times 6'11\frac{1}{4}''$
$57 \times 76\frac{3}{4}$	$15'0'' \times 6'10\frac{3}{4}''$	$15'0\frac{1}{2}'' \times 6'11\frac{1}{4}''$
$33 \times 76\frac{3}{4}$	$11'11'' \times 6'10\frac{3}{4}''$	$11'11\frac{1}{2}'' \times 6'11\frac{1}{4}''$
$45 \times 76\frac{3}{4}$	$15'11'' \times 6'10\frac{3}{4}''$	$15'11\frac{1}{2}'' \times 6'11\frac{1}{4}''$
$57 \times 76\frac{3}{4}$	$19'11'' \times 6'10\frac{3}{4}''$	$19'11\frac{1}{2}'' \times 6'11\frac{1}{4}''$

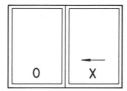 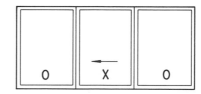

O = FIXED PANEL　　　X = SLIDING PANEL

OUTSIDE LOOKING IN

ALUMINUM SLIDING GLASS DOORS

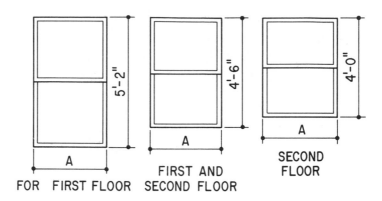

FOR FIRST FLOOR   FIRST AND SECOND FLOOR   SECOND FLOOR

A = IN. INTERVALS OF 2" FROM 1'-4' TO 3'-0"

USUAL SIZES OF WOOD D.H. WINDOWS

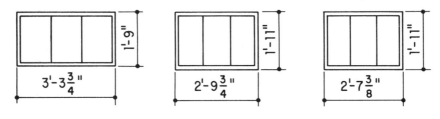

STEEL CELLAR WINDOWS

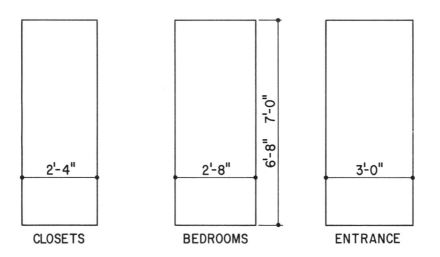

CLOSETS   BEDROOMS   ENTRANCE

STOCK DOOR SIZES

JALOUSIE DOOR

COMBINATION WOOD DOORS

REVIEW QUESTIONS

1. What are the most widely used door types and their available sizes?
2. Name the four window category types based on their use in building construction.
3. Explain what a tabular take-off for doors should contain.
4. What is the cost of a sliding aluminum window for an opening of 3'4" × 4'0"?   What does it cost to install this window?
5. Name four typical window types.

16

PLUMBING

INTRODUCTION

Plumbing is the installation of all piping needed to supply hot and cold water in a building, as well as piping for waste removal. This is generally known as *rough plumbing*, while the installation of fixtures such as the kitchen sink, dishwasher, garbage disposer, lavatory, tub, shower, water closet, water heater, washer, and dryer, including the accessories and fittings for each fixture, is known as *finish plumbing*.

A plumbing estimate consists of the itemizing of all pipe lengths of the various diameters, elbows, tees, Y's, traps, valves, and so on, including the number, type, and size of fixtures—together with material and installation costs.

To accomplish a rough plumbing take-off, the estimator must be thoroughly familiar with the sanitary drainage system as well as the domestic water supply systems. Further, the estimator must understand and be able to read plumbing plans and plumbing riser diagrams.

A description of a typical sanitary drainage and a water distribution system follows.

Sanitary Drainage System

In describing the drainage system it is best to refer to a typical drainage system, such as that shown in Fig. 16-1.

216

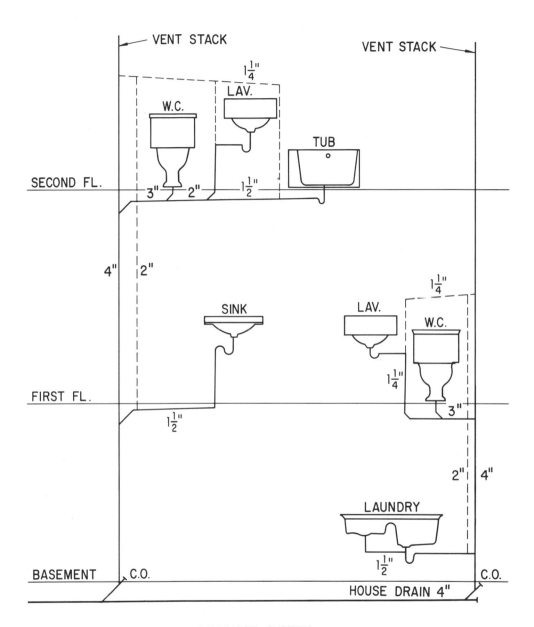

DRAINAGE SYSTEM

Figure 16-1

The *house sewer* is that portion of pipe that extends from the public sewer in the street to about 3 to 5 ft from the house. This pipe may be of hub and spigot, no-hub cast iron, galvanized steel, ABS (acrylonitrile-butadiene-styrene), PVC (polyvinyl chloride), transite, or other approved materials, depending on restrictions of local plumbing codes. ABS and PVC installations are limited to residential construction not more than two stories in height.

The *house drain* is the horizontal pipe inside the house into which vertical soil and waste stacks discharge.

Soil, waste, and vent stacks are the vertical pipes of the drainage system. A distinction must be made between soil, waste, and vent stacks. The soil stacks receive the waste matter from the water closet, while the lavatory and tub drain into a waste stack. The soil stack begins at a point where the water closet branch line connects to the stack. The stack above this point becomes a waste stack if a lavatory, sink, or tub drains into it. The pipe above the point where the lavatory, sink, or tub drains into the stack becomes a vent stack. Vent stacks are used only to allow air to enter the system.

The *fixture branch* is the horizontal pipe leading from the fixture trap to the vertical soil or waste stack. The fixture branch vents are the horizontal pipes that lead from near the fixture trap, then above the fixture to the vent stack. Branch vents are slightly graded so that any condensate that may form can flow back to the branches.

Fixture traps are water seals that prevent odors and gases within the pipe from entering the living area. The trap is directly under the fixture and is U-shaped. The water seal in the trap must not be less than 2 in. and not more than 4 in. except where a deeper seal is found necessary. Traps must be protected from freezing.

Cleanouts are provided on horizontal drainage piping at its upper terminal, and additional cleanouts are required on each run of piping that is more than 100 ft in total developed length, and on a horizontal line for each aggregate change of direction exceeding 135°. Cleanouts must be readily accessible, either above grade or installed under an approved cover plate.

Septic Tank

When there is no public sewer, the house sewer may drain into a private sewer or septic tank (Fig. 16-2), usually constructed of concrete with smooth inside surfaces. The tank retains the solids and digests this organic matter through a period of detention. The liquids in the tank discharge into the soil outside the tank through a system of open-joint piping into a disposal area such as a leaching field (Fig. 16-3).

Leaching field.   The pipe lines of a leaching field can be of clay tile laid with open joints, perforated clay pipe, perforated bituminous fiber pipe, perforated high-density polyethelene, or ABS pipe and PVC pipe. The trench in which the

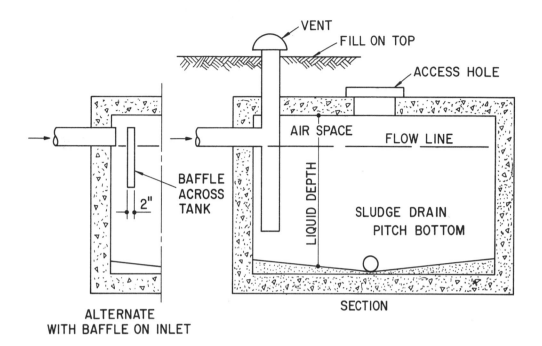

VENT
FILL ON TOP
ACCESS HOLE
AIR SPACE
FLOW LINE
BAFFLE ACROSS TANK
2"
LIQUID DEPTH
SLUDGE DRAIN PITCH BOTTOM
ALTERNATE WITH BAFFLE ON INLET
SECTION

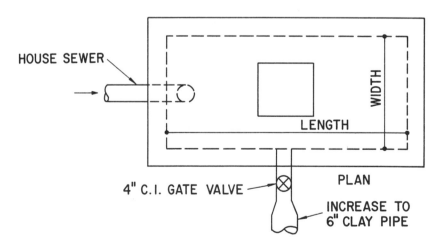

HOUSE SEWER
WIDTH
LENGTH
4" C.I. GATE VALVE
PLAN
INCREASE TO 6" CLAY PIPE

SEPTIC TANK

Figure 16-2

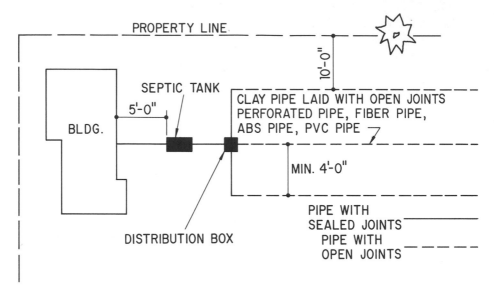

LEACHING FIELD

Figure 16-3

pipe is laid must first have a layer of clean stone, gravel, or slag varying in size from $\frac{3}{4}$ to $2\frac{1}{2}$ in.   Drain lines must be covered with untreated building paper, straw, or similar porous material to prevent closure of voids in the pipe with earth backfill.

Domestic Water Supply System

The piping system in Fig. 16-4 is that of a water distribution system in a multistory frame building.   The water comes from a public waterworks where it is purified and sent under pressure through pipes to the water service line that runs from the street to the house.   The water is delivered at pressures ranging from 55 or 60 to 75 lb per square inch (psi), and may be as high as 100 psi.

High water pressure can eventually damage faucets and fittings.   Water hammer, a banging noise caused by vibrating piping, can occur when a valve abruptly stops the pressurized flow of incoming water.   Water hammer can occur even if a high-speed wall of liquid, as from a starting pump, hits a sudden change of direction in the piping, such as an elbow.

This problem can be eliminated by installing ready-made shock absorbers, or by assembling an air chamber from standard pipe and fittings over lavatories and tubs.   By installing a pressure-reducing valve in the service line near the point where the water enters the house, the line pressure can be controlled down to 60, 55, or 40 psi.   Turn the nut on the pressure valve until the gauge shows the desired pressure.   A water pressure gauge can be purchased from your local plumbing

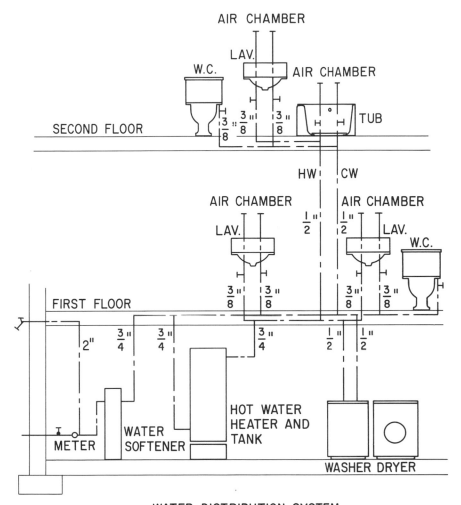

WATER DISTRIBUTION SYSTEM

Figure 16-4

supply store.   Screw the gauge onto the sill cock and open the faucet and read the water pressure in pounds per square inch.   *Caution:* Make certain that all other faucets are turned off during this pressure testing.

Domestic Water Pipe

The piping used for the fresh water system may be of copper, plastic, wrought iron, or galvanized steel.   Copper joints are sweated and plastic pipe joints are assembled with a liquid glue, while steel and wrought iron pipe is threaded.

Copper is available as type L pipe of medium weight, used for interior work. It is also available as type K tubing, a heavier pipe used for underground work, which comes in soft temper in 20-ft straight lengths, or in coils in sizes from $\frac{3}{8}$ to 6 in. in diameter. Copper tubing is also manufactured in type M, lightweight, used for carrying water under pressure, and type DWV (drain–waste–vent), thinner than type M—used for drain, waste, and vent lines where permitted, but not for water. ARC tubing is manufactured for air-conditioning and refrigeration.

According to most plumbing codes, gas piping must be of wrought iron or steel (galvanized or black). In some areas PVC pipe may be used in exterior buried piping systems. Wrought-iron and steel gas piping must have threaded joints.

THE TAKE-OFF

A method of taking-off pipe lengths, fittings, and fixtures from the plans and riser diagram (Fig. 16-5) is shown in tabular form in Fig. 16-6. The take-off was started with all vertical piping on the riser diagram, beginning with the 4-in. vent extending 4 ft above the roof. The 4- to 2-in. reducer coupling connects the 2-in. vent stack, 21 ft long. This is followed by a 2- to 4-in. Y-fitting, then another Y-fitting, a length of 4-in. soil stack, and so on.

This take-off method lists all pipe lengths and fittings in the order in which they appear on the diagram. If the same-size fitting appears several times, the quantity number under the column heading "number of fittings" can be corrected.

When all vertical drain and vent lines and fittings on the riser diagram are listed, take-off the pipe line lengths and fittings on the plans. Remember that horizontal lines on the riser diagram are not true lengths and should not be scaled. Only horizontal lines on the plans can be scaled. Exact scaling, however, is impossible and is not even desirable, but this method is close enough for all practical purposes.

Finally, the fixtures are counted and listed.

Material unit and installation costs can be gotten from current publications lists, or locally where the job is done.

Domestic Water System Take-Off

Piping for domestic water, both hot and cold, is estimated from water distribution plans and riser diagrams, similar to that for the sanitary system. Pipe lengths of the same diameter are listed, together with fittings and valves. The take-off may begin either at the end or the beginning of the pipe run, listing fittings and valves as they occur.

Unit and installation cost prices of pipe lengths, together with the costs of fittings and valves, can be found in current publications.

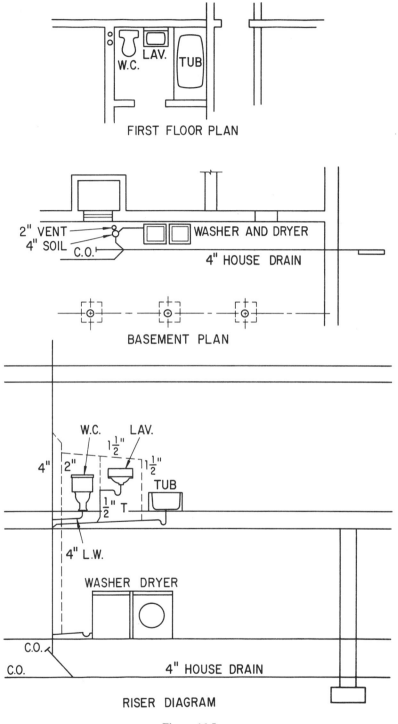

FIRST FLOOR PLAN

2" VENT
4" SOIL
C.O.

WASHER AND DRYER

4" HOUSE DRAIN

BASEMENT PLAN

W.C.   LAV.
$1\frac{1}{2}$"
4"   2" $1\frac{1}{2}$"
TUB
$1\frac{1}{2}$" T
4" L.W.

WASHER DRYER

C.O.
C.O.   4" HOUSE DRAIN

RISER DIAGRAM

Figure 16-5

223

TAKE-OFFS FROM PLUMBING PLANS AND RISER DIAGRAM (REFER TO FIG. 16-5)					
Item	Mat'l Unit Cost	Install. Unit Cost	Mat'l and Install. Cost	Linear Feet of Pipe	Number of Fittings	Total Cost
From Riser Diagram						
All Vertical Drainpipe and Fittings						
4" Vent from roof	4.89	7.20	12.09	4		48.36
Reducer-coupling 4" TO 2"	3.00	1.79	4.79		1	4.79
2" Vent stack C.I.	2.74	6.35	9.09	21		190.89
2" Y-fitting C.I.	6.60	7.05	13.65		1	13.65
2" to 4" Y-fitting C.I.	8.20	9.90	18.10		2	36.20
$1\frac{1}{2}$" to 4" Y-fitting C.I.	15.40	12.70	28.10		1	28.10
4" Soil stack C.I.	4.69	7.20	11.89	20		237.80
4" Cleanout C.I.	48.00	29.00	77.00		1	77.00
4" Y-fitting C.I.	17.60	12.70	30.30		1	30.30
						667.09
Vertical Vent Piping and Fittings						
$1\frac{1}{2}$" Fixture vents	12.45	4.54	16.99	39		662.61
3" Pipe vent	15.55	4.38	19.93	4		79.72
$1\frac{1}{2}$" T-fitting	8.25	8.00	16.25		4	65.00
$1\frac{1}{2}$" 90° elbow	6.60	5.35	11.95		3	35.85
$1\frac{1}{2}$" Trap	24.00	23.00	47.00		3	141.00
						984.18
Fixtures						
W.C. 2-pc. tank, china	245.00	60.00	305.00		1	305.00
Lav. with vanity top	85.00	50.00	135.00		1	135.00
42"x 37" tub and shower	410.00	54.00	447.00		1	447.00
Washer, automatic	340.00	59.00	399.00		1	399.00
Dryer	260.00	91.00	351.00		1	351.00
Sink, Kitchen, 2-bowl	145.00	120.00	360.00		1	360.00
Water softener	320.00	71.00	391.00		1	391.00
						2388.00
					Grand total = $	4039.27

Figure 16-6

MATERIAL AND INSTALLATION COSTS

The costs for fixtures are:

Water closet: A two-piece water closet, floor-mounted, can be installed in 2 hours. A white ceramic water closet costs between $125 and $150.

Lavatory: vanity top, with fittings, white porcelain enamel on cast iron, costs between $130 and $160.

Bath tub: white porcelain enamel on cast iron costs between $550 and $740. Preformed fiberglass or other plastic preformed units cost about $300. A shower mixing valve can be installed in 1 hr and costs between $130 and $150.

Kitchen sink: counter top, porcelain enamel on cast iron, 32 in. × 20 in. double bowl, costs about $260 in place. A two-person team requires about 3 hr to install a unit.

Dishwasher: built-in, four-cycle, costs between $400 and $450.

Garbage disposer: sink type costs between $110 and $140.

Washing machine: automatic costs between $275 and $695.

Dryer: automatic, costs between $280 and $525.

Water heater: residential, glass lined and gas fired, 40-gal capacity, costs about $200 to $225. One plumber can install the tank in about 4 hr.

Septic tank: 2000-gal tank costs $750 installed, not including excavation or piping.

Water softener: automatic, costs $455 installed.

The costs for piping are:

Galvanized steel pipe: from $\frac{1}{2}$- to 3-in. diameter, with threaded joints, costs about $4.25 per linear foot. 65 to 75 lin ft can be installed per 8-hr day by one plumber.

Copper pipe: $\frac{1}{4}$-in. diameter, with solder joints, costs about $4.10 per linear foot. 50 to 60 lin ft can be installed per 8-hr day by one plumber.

PVC (polyvinyl chloride): $\frac{1}{4}$-in.-diameter pipe costs about $4.20 per linear foot. 45 to 50 lin ft can be installed per 8-hr day by one plumber.

ABS (acrylonitrile-butadiene-styrene): plastic pipe, $1\frac{1}{2}$-in. diameter, costs $7.40 per linear foot. 27 to 30 lin ft can be installed per 8-hr day by one plumber. 4-in.-diameter pipe costs $12.35 per linear foot installed.

CPVC: plastic pipe for hot water line, $\frac{1}{2}$-in.-diameter, costs $5.30 per linear foot. 47 to 50 lin ft can be installed per 8-hr day by one plumber.

SELF EXAMINATION

1. Find the cost of the two vertical plumbing stacks and the stack fittings shown in Fig. 16-1. Omit all vents and branch lines. Assume that both stacks and fittings are of ABS material and that each stack length is 29 ft.

2. Find the total cost of fixtures. Refer to Fig. 16-6 for fixture prices.

Answer Box	
Plumbing stacks	$1285
Fixtures	$2923

ASSIGNMENT—16

1. In Fig. 16-4, estimate *only* the linear feet of vertical cold water piping (approximately), including 24-in.-high air chambers. Cold water piping is $\frac{3}{8}$ in., $\frac{1}{2}$ in., $\frac{3}{4}$ in., and 2 in. Assume piping to be PVC plastic pipe. Find the linear feet of piping of each diameter.

2. Find the cost of the piping.

SUPPLEMENTARY INFORMATION

Air Chambers

Water hammer in the domestic water supply system can be eliminated by installing air chambers. These are small lengths of vertical pipe about 18 in. long, capped at the top and attached to the line near a valve or faucet. The air chamber contains air which gets compressed when the water hammer wave hits it and acts as a cushion that greatly reduces the sound and impact.

When water hammer starts on a system that has been quiet before, it becomes quite obvious that the air chambers have filled with water, making them ineffective. To remedy this, close off the supply from the main water valve and open all faucets to drain out the water, thereby allowing air to enter the piping system. The air chambers will fill with air once again and become effective. Then close all faucets and turn on the main water valve.

Water Softener

Water softeners are designed to remove large quantities of calcium and magnesium or bicarbonates that are contained in hard water. Hard water may also contain objectionable minerals, such as iron and sulfur. Its constant use as drinking water may cause intestinal disorders, and its flow through piping will cause the formation of scale, adding resistance to the flow of water.

Calcium and magnesium react with soaps, taking away the cleaning power of the soap. Iron in hard water may cause stains on clothing, while sulfur creates an objectionable odor and taste to the water.

A mineral known as zeolite is put into the water softener tank to exchange the calcium and magnesium for sodium (salt), which reacts favorably with soap. From time to time the sodium must be replaced by adding common salt to the softener. Drinking water treated in this manner has a salty taste and should be taken in moderation for those on a low-salt diet.

Storage Capacities of Water Heaters

Storage capacities of water heaters for residences range from 20 to 60 gal.

Minimum	20 gal
Family of two	30 gal
Family of four	40 gal
with dishwasher	
and automatic washer	50 gal
Larger families	60 gal

Gas Supply Piping

Furnace inputs up to 125,000 Btu/hr generally do not require a gas supply pipe size larger than 1 in. However, in no instance must a gas supply line be smaller than $\frac{1}{2}$ in., and a gas supply line must not serve more than one unit.

REVIEW QUESTIONS

1. On the sanitary drainage system, the house sewer piping can be of what material? Name three.
2. In estimating fixture–branch line piping, can the lengths of pipe be scaled from the plans or the riser diagram?
3. From your local plumbing supplier, get the unit and installation costs of the following:
   (a) Lavatory—single bowl, vanity top.
   (b) Water closet with tank.
   (c) Dishwasher.

17

HEATING, VENTILATING, AND AIR-CONDITIONING

INTRODUCTION

In this unit you will learn the basic concepts of heating, ventilating, and air-conditioning, generally referred to as HVAC, together with a take-off of sheet metal for a duct system.

Of the many types of heating systems in use, we will concern ourselves with two of the most commonly used systems: systems where hot water is distributed through a system of pipes, known as hot water systems, and systems in which heated air is circulated through ducts into rooms to be heated, warm air systems.

Hot Water System

Of the hot water systems, the one-pipe reversed return is perhaps the most popular (Fig. 17-1). The hot water coming from the boiler into the first radiator also feeds the second and third, and so on. The return line from each radiator, containing somewhat cooler water, is collected and brought back in one main line to the boiler.

The first radiator has the shortest supply main but the longest return main. For the farthest radiator, the reverse is true. No matter where the radiator is positioned, the total length of its supply and return mains are essentially equal to those for every other radiator. This guarantees delivery of an equal amount of heat in each radiator.

Other hot water heating systems are the two-pipe direct return, the series-

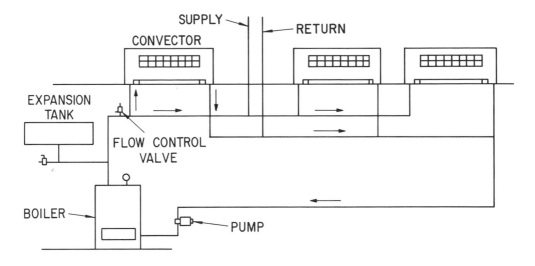

TWO-PIPE REVERSED RETURN HOT WATER SYSTEM

Figure 17-1

loop hot water systems of the one-pipe single-circuit system, and the one-pipe two-circuit system.

Warm Air Duct System

The warm air duct system (Fig. 17-2) is one of the most popular types of residential heating systems today. Warm humidified air is forced through the system by a blower and distributed in measured amounts through directional outlets in each room.

The duct system is arranged to provide for the admission of outside air for ventilation in an amount equal to about 20% of the total air circulated in the winter months, but may also permit circulation of 100% outdoor air in the spring, summer, or fall. In addition, the air is filtered at all times by inexpensive, throwaway filters located in the blower section adjacent to the furnace. A humidifier pan, automatically filled with city water, is located in the furnace unit and has sufficient capacity to maintain a relative humidity of about 50% in the heated area.

The gas or oil burner, motor, blower, filters, and humidifier are all contained in an insulated metal jacket, giving the unit an acceptable appearance as part of a utility or recreation room.

The forced warm air system for the average-size home is capable of supplying between four and six complete air changes per hour within the heated space. Warm, filtered, and humidified air leaves the heating unit at about 155°F and

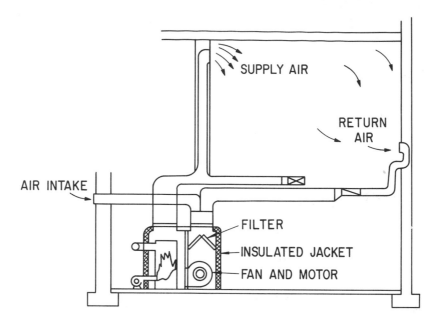

FORCED WARM AIR SYSTEM

Figure 17-2

arrives at the room grille at about 140°F, having lost some of its heat through the metal ductwork above the ceiling, or in partition walls.

The matter of whether to install the supply grille high on the side wall or near the floor depends on the climate of the area in which the house is located.

In northern and colder climates, where cold floors can be a problem and where walls facing outside exposures must be thoroughly heated, supply air diffusers are best placed around the perimeter of outside-facing walls. The air diffuser can be located near the baseboard.

In warmer climates, however, where cooling air distribution is more important than heating air distribution, diffusers can be located on an inside wall as high as 6 ft above the floor or about 6 in. from the ceiling. In this location the outlets are high enough to blow air over the heads of occupants and are also ideally placed for cool air supply should summer air-conditioning be added to the system in the future. Adjustable bars across the face of registers direct the air in any desired direction.

In addition to one or more supply outlets, each area except the kitchen and bathrooms contains a return-air outlet, for the purpose of recirculating the returned 80% of room air. No special construction is required where ducts run through partitions, as the ducts are sized to fit the standard stud wall space. For rooms larger than 10 ft × 12 ft, two supply registers will be satisfactory.

Where one or more supply outlets are placed in every room to be heated, connecting rooms with open doorways can have a common return that may be located in a hallway or foyer. This is particularly true of smaller homes of 1500 sq ft or less.

The furnace unit may be placed wherever convenient within the house or in an attached garage. Ducts may be run in the basement or in the attic space, or along the ceiling of the room being heated. If ducts are run above the ceiling, air outlets of special design may be mounted on the ceiling.

The temperature control system for a warm air heating system is somewhat similar to that used on a hot water installation. The room thermostat, upon demand for heat, starts the blower. But the blower can run only if the air temperature at the furnace outlet is between 90 and 175°F. This prevents the circulation of air which is too cool to heat the room. If the air temperature at the furnace outlet drops below 90°F, the burner operates to heat the air to at least 90°F before the fan is allowed to start. When the room thermostat is satisfied, the blower and burner units do not operate.

The advantages of the warm air duct system can be stated as follows:

1. Provides all the requirements of winter air conditioning, air cleaning, and air circulation.
2. May be converted into year-round air-conditioning by adding summer cooling equipment.
3. System cannot freeze in the winter.
4. Can circulate air without cooling or heating.

Air-Conditioning

When the warm air duct system is converted into year-round air-conditioning, it requires the installation of a condenser and a compressor unit on a concrete pad outside the building with the heating–cooling blower section inside the building. The air-conditioned air is then distributed throughout the building by a duct system, controlled with dampers, splitters, grilles, or registers. The operation is controlled automatically by thermostats.

Air-Conditioning Calculation

Before any cost figures for heating or air-conditioning can be found, it is necessary, first, to arrive at the requirements obtained by calculating the heat loss or heat gain of the building. To understand and do this, the heat brought into a room by a heating system is constantly being lost through walls, windows, ceilings, and floors when there exists a temperature difference between the inside and outside of the walls, windows, ceilings, and floors.

The amount of heat that will be lost depends on (1) how much surface is exposed to the colder temperature, (2) how effective a heat barrier (insulation) the surface is, and (3) how great the temperature difference is that is causing the heat flow.

A simple rule for computing heat loss is

heat loss = area × wall effectiveness × degree temperature difference

where heat loss is measured in Btu (British thermal units) per hour. "Area" refers to the square feet of only those surfaces that are exposed to a temperature less than room temperatures. By "surface" is meant walls, glass, door, ceiling, floor, and partition. No heat loss occurs between adjacent rooms when each room is heated to the same temperature.

"Wall effectiveness" expresses the ability of a material to resist the flow of heat. A heat-flow transmission coefficient has been assigned to every type of commercially used construction. Constructions with low coefficient factors indicate good insulators, while those with relatively high factors indicate poor insulators.

These factors are called U-factors, and have been derived by the American Society of Heating, Refrigeration and Air-Conditioning Engineers (ASHRAE). U-factors for residential construction vary from 0.07 for well-insulated walls to 1.13 for window glass and thin doors.

The U-factor represents the amount of heat in Btu that will pass through 1 sq ft of a surface per hour per degree Fahrenheit of temperature difference between the outside and inside air temperatures. The U-factor is also the reciprocal of the sum of the thermal resistance values (R) of each element of a structural section.

The materials of a masonry cavity wall, with its elements of 4-in. face brick, an air space, 4-in. common brick, and gypsum lath with a $\frac{1}{2}$-in. plaster finish on the inside, have thermal resistance values as shown in the following:

	R-Value
Outside surface air (15-mph wind)	0.17
Face brick, 4 in.	0.44
Common brick, 4 in.	0.80
Air space	0.97
Gypsum lath, $\frac{3}{8}$ in.	0.32
Plaster–sand aggregate, $\frac{1}{2}$ in.	0.09
Inside surface (still air)	0.68
Total resistance	3.47

Note: The U-factor is also the reciprocal of the sum of the thermal resistance values (R) of each element of the structural section.

$$U = \frac{1}{R} = \frac{1}{3.47} = 0.29$$

The temperature difference (Td) is the arithmetic difference between the inside room temperature and the coldest outside temperature likely to occur in a given locality expressed in degrees Fahrenheit. It is common practice to use 70°F for inside design temperature, with the outside temperature varying with geographical locations. The outside design temperature is not the coldest temperature ever recorded in a given locality, since extreme low temperatures exist for only short periods of time and it would not be economical to design for such a condition.

Following are but a few outside design temperatures in a few parts of the country:

California
    Los Angeles   40°
    San Francisco   35°
Illinois
    Chicago   −10°
    Springfield   − 5°
New Jersey
    Newark   5°
    Trenton   10°
New York
    Buffalo   0°
    New York City   10°

The basic rule for heat loss can be stated as follows:

$$\text{heat loss in Btu per hour} = A \times U \times Td$$

Infiltration

In any room containing windows and doors, there is a certain amount of air that seeps in through window and door cracks when the wind blows. This leakage is called infiltration. Every cubic foot of cold outside air that leaks into the room imposes an additional load on the heating system. It is entirely correct to consider the heat required to warm the infiltrated air to room temperature as an additional heat loss measured as the heating system has to supply it.

The method used in calculating the air infiltration is to find the cubic content of the room and multiply the volume by the U-factor 0.018 by the design temperature difference.

As an example, a heat-loss calculation for a single corner room (Fig. 17-3) of a small building is shown below as it would be recorded on a worksheet.

Ductwork for Warm Air Heating

Air entering a room at 115°F will give up (115°F − 70°F room temperature) = 45°F of its temperature in replacing the room heat loss to maintain the design room

HEAT-LOSS CALCULATION

| Name: Corner Room | | | | Design Temperature: Outside | | 5°F | |
| Location: Newark, N.J. | | | | | Inside | 70°F | |

Room size	Item	A	U^a	Td	Btu/hr	Total Btu/hr
30′ × 20′ × 8′ = 4800 cu ft	Exposed wall 30 + 20 = 50 × 8	400				
	Windows 4 × 3 × 4 = 48	− 48	1.12	65	3,494	
	Net wall	352	0.28	65	6,406	
	Ceiling 30 × 20 =	600	0.24	65	9,360	
	Floor 30 × 20 =	600	0.26	65	10,140	
					29,400	
	Infiltration = 4800 × 0.018 × 65 = 5616				5,616	
					35,016	35,016

ᵃFor U-factor, refer to ASHRAE.

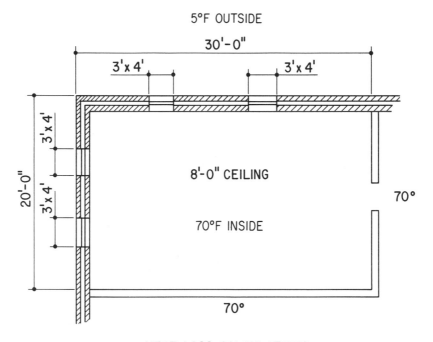

HEAT-LOSS CALCULATIONS

Figure 17-3

temperature of 70°F.   This process requires a calculated amount of air as follows:

$$\text{cfm} = \frac{\text{heat loss, Btu/hr}}{Td \times 1.08}$$

where cfm = cubic feet of air per minute; Td = temperature difference between entering room air temperature and room temperature; and 1.08 = a constant derived from the specific heat of air (0.241 Btu/lb dry air), 60 (minutes in 1 hr), and 0.075 (density of air, lb/cu ft at sea level).   Thus

$$0.241 \times 60 \times 0.075 = 1.08$$

Furnace Heating Capacity

This is the sum of the heat loss (Btu/hr) plus the heat required to warm the fresh air being introduced (not the same as infiltration), plus a pickup loss allowance for warming a building on a cold day after a shutdown period.   This pickup factor is usually about 60%, or 1.6 multiplier.   Therefore,

$$\text{furnace capacity} = \text{room heat loss} + \text{pickup factor}$$

The fresh-air load can be found by a cross-multiplying the cfm formula and using Td = (room temperature − outside air design temperature), such as

$$\text{cfm} = \frac{\text{heat loss (Btu/hr)}}{Td \times 1.08}$$

Cross-multiply:

$$\text{heat loss} = \text{cfm} \times 1.08$$

where cfm = fresh air being used (usually 10% of supply cfm)

$$Td = 70°F - 0°F$$

The 0°F is the design outside temperature in Buffalo, New York, but this can change for other areas.   The outside design temperature for San Francisco, California, is 35°F.

Data on Sizing a Duct System

Ductwork of an air-conditioning or ventilating system is often drawn in single-line form, as shown in Fig. 17-4.   This is particularly true in large buildings, where the usual plans drawn to a $\frac{1}{4}$-in. scale would be too large to show the entire plan on one sheet.   A $\frac{1}{8}$-in. scale drawing would then be used, provided that the duct layouts also permit space on the plans for plumbing, heating, and other pipe lines.

Every duct system has a main duct, branch ducts, and register outlets.   The main duct starting at the unit is usually of the largest cross-sectional size and diminishes in size as air volume is given off into the branch lines.

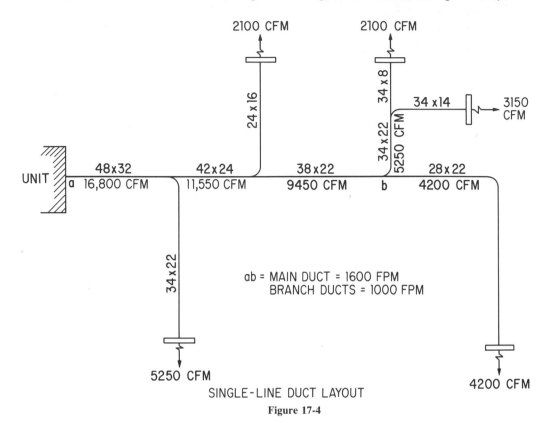

SINGLE-LINE DUCT LAYOUT

Figure 17-4

In determining the size of ducts in a duct system, the heated air (cfm) that is to be delivered into a room must be known, as discussed previously. Next, the velocity of the air, in feet per minute (fpm) must be known. With these two factors, the cross-sectional area in square feet can be found by the formula

$$A = \frac{\text{cfm}}{\text{fpm}}$$

Some of the accepted air velocities through ducts in residences and for public buildings are as follows:

	Low-velocity system (fpm)	
Description	Residences	Public buildings
Main ducts	500	1200–1600
Branch ducts	450	600–1000
Wall stacks	350	400–600
Baseboard registers	300	100–400
Wall registers above 5 ft	500	300–500
Outside air intakes	—	1000

Sizing Ducts

From the formula A = cfm/fpm, the 28 in. $\times$ 22 in. duct carrying 4200 cfm was found thus:

$$A = \frac{\text{cfm}}{\text{fpm}}$$

$$= \frac{4200}{1000} = 4.2 \text{ sq ft} \times 144 = 604.8 \text{ sq in.}$$

604.8 sq in. translates into a 28 $\times$ 22 in. duct.

THE TAKE-OFF

Heating and air-conditioning costs vary so much that it is necessary in determining these costs to rely on a contractor who specializes in these fields. However, since duct work plays such an important role in warm air heating, ventilating, and air-conditioning, we will estimate take-offs to the sheet metal used on the duct system.

Sheet metal is figured by the weight of material used and the time required to fabricate and install the metal. Following are but a few commonly used sheet metal gauges used in ductwork.

SHEET METAL GAUGES

Gauge number	Thickness (in.)	Weight per square foot (lb)
24	0.025	1.000
25	0.0218	0.875
26	0.0187	0.75

Estimating Duct Work

The quantity of sheet metal required for a duct system is estimated from the typical shop drawings. Each joint or fitting is itemized or identified by a circled number. The length of each joint or fitting is multiplied by its perimeter to find the number of square feet, which is then multiplied by the weight of the metal per square foot of the metal gauge selected. An allowance of 35% is made for waste in cutting.

For example, estimate the weight of the metal required for the duct partial shop drawing shown in Fig. 17-5. The metal gauge used is 26.

Material and installation costs for galvanized steel are $4.32 per pound. Therefore,

$$4.32 \times 263 = \$1136.16$$

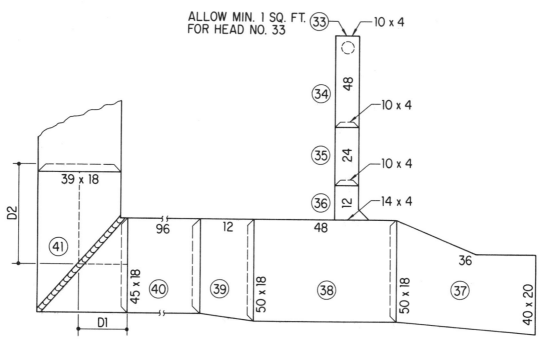

NOTE: LINE D1 + D2 = LINEAR FEET.
ALWAYS USE LARGER PERIMETER. IGNORE SLIGHT
BEND. CALCULATE AS STRAIGHT RUN.

Figure 17-5

TAKE-OFF QUANTITIES FROM SHOP DRAWINGS (Fig. 17-5)

Piece number	Duct size (in.)	Linear footage	Perimeter (ft)	26-Gauge sheet metal	
33	Head			1	
34	Head			12	sq ft
35	10 × 4	4	3 min.	6	
36	10 × 4	2	3 min.	3	
37	14 × 4	1	3	34	
38	50 × 18	3	11.33	45	
39	50 × 18	4	11.33	11	
40	50 × 18	1	11.33	84	
41	45 × 18	8	10.5	63	
	45 × 18	6	10.5	259	
				× 0.75	lb
				194.25	
				× 1.35	
				262.23	or 263
			Plus 35% waste		lb

MATERIAL AND INSTALLATION COSTS

Galvanized steel: for ductwork and installation costs

Under 450 lb: $4.32 per pound

450 to 1000 lb: $3.61 per pound

Fan for air-conditioning:

15,600 cfm, 10 hp $3600

3,800 cfm, 5 hp $2045

210 cfm $136

Grilles: The installed costs are:

6 in. × 6 in.: $12.40

10 in. × 6 in.: $13.45

24 in. × 18 in.: $31.80

Furnace: for hot air heating, including blower and standard controls completely installed:

47,000 Btu/hr = $510

76,000 Btu/hr = $605

Solar energy system: not including connecting pipe, plumbing, and so on:

two collectors, circulator, fittings, 120-gal tank: $2045

three collectors, circulator, fittings, 120-gal tank: $2575

SELF EXAMINATION

Work out the following problem and check your answers with those given in the Answer Box.

1. Estimate the quantity or weight of the main duct (Fig. 17-6). Use 26-gauge metal.
2. Find the material and installation costs.

Answer Box
Weight of material = 128.58 lb Material and installation costs = $555.50

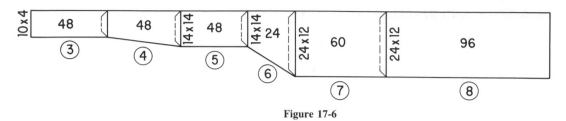

Figure 17-6

ASSIGNMENT—17

1. Estimate the weight of the material of the single-line duct system shown in Fig. 17-4. The metal gauge is 26.
2. Estimate the cost of the material.

SUPPLEMENTARY INFORMATION

Ventilation

Mechanical ventilation can be achieved through the use of fans and blowers. First, find the volume of the space to be ventilated. Second, using the suggested air-change rates shown on the following chart, find the selection that best meets the condition of the air space to be ventilated.

$$\text{cfm} = \frac{\text{air-space volume in cubic feet}}{\text{air-change rate}}$$

SUGGESTED AIR-CHANGE RATES

Type of building	Minutes per air change
Assembly hall	5–10
Classroom	4–6
Dining room	10–15
Kitchen	1–3
Residence	2–6
Toilet	2–5

For example: for a dining room 30 ft × 40 ft and 20 ft high, find the cfm to properly ventilate the room.

$$\text{cfm} = \frac{30 \times 40 \times 20}{15} = 1600$$

A fan size with this capability can be selected.

REVIEW QUESTIONS

1. What is the Btu/hr heat loss through an ordinary window whose U-factor is 1.13 when there is a temperature difference of 65°F?   The window is $3'6'' \times 4'8''$.

2. What is the R-value of the 1.13 U-factor?

3. In the formula

$$\text{cfm} = \frac{\text{heat loss (Btu/hr)}}{Td \times 1.08}$$

what does the 1.08 represent?

18

ELECTRICITY

INTRODUCTION

Electrical work, like plumbing, is highly specialized, and is estimated and installed by the electrical subcontractor.

Similar to plumbing, electrical work can be divided into two phases: that which is hidden in floors, ceilings, and walls, such as conduits for wiring, boxes for outlets, and so on; and finished work, such as lighting fixtures, switches, fans, motors, and other equipment.

On floor plans, electrical work is generally shown by symbols representing outlets, switches, and fixtures. The estimator must be thoroughly familiar with such symbols and circuits before a take-off can be made.

Electricity is delivered to the consumer at a pressure known as voltage. Lamps, appliances, and other electric power-driven devices are made to operate at a specific voltage of 110/120 volts (V) and 220/240 V.

Water pipes that are corroded on the inside cannot deliver the proper amount of water. Similarly, electric wire that is too small in diameter for the current reduces the voltage (pressure) and reduces the efficiency of lamps, appliances, and other electric-driven power devices.

Electric wire is the conductor of electricity and there are different wire sizes used for different amounts of current. Current is measured in amperes (A). A small-diameter wire cannot carry as many amperes as can a large-diameter wire.

Wire sizes are indicated by gauge numbers. In house wiring the gauges range from 0 to 14, where the 0 number represents the larger diameter. Electric wire is usually of copper, since it is an excellent conductor of electricity. The wire size used in residential work is No. 14. A No. 12 size may be specified in some instances. Electric kitchen ranges require No. 10 or No. 8 wire, and powered equipment may demand a wire as large as No. 2 or No. 0.

Wire installation. Most electric wires are wrapped in some form of electric insulation, which is marked with letters representing the kind of wrapping material used. For example:

R: rubber insulation and cotton covering.
RH: same as R but with greater resistance to heat
RW: same as R but with moisture-resistant rubber and corrosion.
AC: armored cable

Conduits are the pipes in which one to five or more electric wires are run. Conduit pipe is similar to water pipe except that the metal is softer, for easy bending. It is, however, rigid and rust resistant.

Electric circuit. When we speak of an electric circuit, terms such as rate of flow, pressure, and resistance need to be clarified. In electricity, the rate of flow is measured in amperes (A), the pressure in volts (V), and the resistance in ohms (Ω).

The rate of flow (amperes) equals the pressure (volts) divided by the resistance (ohms). This can be stated simply as one of the following:

$$\text{amperes} = \text{volts} \div \text{ohms}$$

$$\text{volts} = \text{amperes} \times \text{ohms}$$

$$\text{ohms} = \text{volts} \div \text{amperes}$$

A further simplification is possible by using letters to represent the above:

$$I = \text{rate of flow (amperes)}$$

$$E = \text{pressure (volts)}$$

$$R = \text{resistance (ohms)}$$

Then we can say that

$$I = \frac{E}{R}$$

$$E = R \times I$$

$$R = \frac{E}{I}$$

Watts. There is another unit of measurement used in electricity besides the volt, ampere, and ohm. This is the watt (W). A watt is a unit of electric power. All lamps, toasters, motors, and other electrical equipment have the number of watts they consume marked on them. One thousand watts is 1 kilowatt (kW). One horsepower (hp) is equal to 746 W, and $\frac{1}{6}$ hp is $746 \div \frac{1}{6} = 128\frac{2}{3}$ W.

The number of watts that an electric circuit can safely take is found by multiplying the amperes by the number of volts. Another way to say this is

$$watts = amperes \times volts$$

An electric circuit rated at 15 A at 120 V can safely take $15 \times 20 = 1800$ W, or 1.8 kW.

From the above it can be seen that the flow of an electric current is proportioned to the pressure imposed upon the circuit and the resistance of the circuit.

The service entrance conductor consists of wires which carry the current from the power company lines into the building. The service entrance conductor may be underground in conduit pipes or be strung overhead from a street pole to the building.

These wires must be of a size that meet the electrical code requirements, and nothing smaller than a No. 10 gauge up to 50 ft in length is permitted. A No. 8 gauge wire is recommended to withstand wind and ice and must clear the ground by at least 18 ft over driveways and 10 ft over walks.

To find the correct size of service entrance wire from pole to the house, the distance from the pole to the building must be known plus the amperes.

SERVICE ENTRANCE WIRE SIZE

Load in building (A)	Distance from pole to building (ft)	Recommended wire size number
Up to 25 A, 120 V	Up to 50	10
	50–80	8
	80–125	6
20 to 30 A, 240 V	Up to 80	10
	80–125	8
	125–200	6
	200–350	4
30 to 50 A, 240 V	Up to 90	8
	80–125	6
	125–200	4
	200–300	2
	300–400	1

The electric service for new homes up to 3000 sq ft is at least 100 A. This takes care of the many new appliances that are now used. This brings the electric service up to 10,000 W, requiring a No. 2 or No. 3 wire with RH insulation.

Reading the Riser Diagram

The riser diagram shown in Fig. 18-1 is that of a two-story library building with cellar and roof electrical circuits and equipment. From the main distribution panel, in the cellar, which houses fuses or circuit breakers and the main switch, feeder lines of No. 4 wire with RH insulation and housed in $1\frac{1}{2}$-in.-diameter pipe conduits are led to lighting panels in the cellar, first floor, and second floor. The lighting panels are indicated as LPC, LP1, and LP2, respectively. The lighting panels are usually located in a convenient place and as close as possible to where the greatest electrical load is required.

From the lighting panel run the various circuits for lighting outlets, fan outlets, and other receptacles. LPC radiates 16 circuits, LP1 radiates 20 circuits, and LP2 radiates 13 circuits.

Two other lines from the main distribution panel serve a $\frac{3}{4}$-hp motor for a lecture room fan, plus two $\frac{1}{6}$-hp motors on the roof serving a lecture room exhaust fan and toilet No. 2 in the cellar.

Types of circuits. In distributing electrical energy throughout a building, three types of circuits may be employed:

1. *General-purpose circuit:* used for lighting outlets and convenience outlets. It is usually designed for one 20-A, 120-V circuit for not more than 500 ft of floor space. The wattage for this circuit is usually up to 2400 W.
2. *Appliance circuit:* serves appliances used in the kitchen, such as refrigerator, toaster, and other electrical conveniences. This circuit should have at least two 20-A, 120-V circuits. The wattage for one circuit may be 2400 W.
3. *Individual equipment circuit:* supplies single electrical appliances of higher wattage, such as an electric range, which can consume 800 W or more.

An 800-W range at 120 V requires how many amperes?

$$\text{Amperes} = \text{watts} \div \text{volts}$$
$$= 800 \div 120$$
$$= 66.6$$

Another individual equipment circuit is that which feeds a central air-conditioning system at 5000 V. This circuit requires 240 W. How many amperes does this circuit require?

$$\text{Amperes} = \text{watts} \div \text{volts}$$
$$= 5000 \div 240$$
$$= 20.8$$

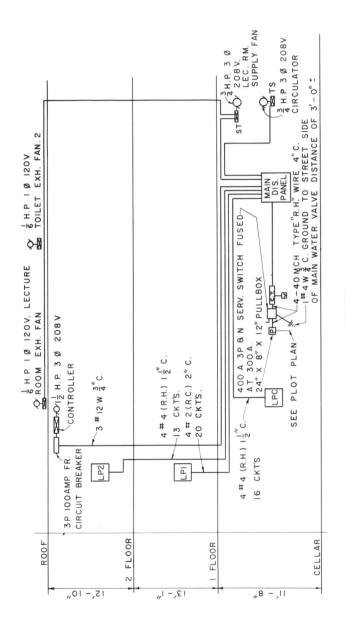

RISER DIAGRAM

NOT TO SCALE

LIGHT & POWER

Figure 18-1

246

Low-Voltage Riser Diagram

The low-voltage circuit shown in Fig. 18-2, serving bells, clocks, chimes, and buzzers, receives current from circuit 1A-21. A transformer T is used reducing the voltage down to about 6 to 24 V from the 110/120-V circuit. After the transformer, the circuit number becomes 1A-28, and serves a desk pushbutton pad, followed by the low-tension wiring to chimes, bells, buzzers, and clocks.

Reading the Electrical Plan

The partial first-floor electrical plan (Fig. 18-3) is that of the two-story library building with cellar. The numerous curved lines representing wiring to lighting fixtures, duplex receptacles, floor outlets, switches, and so on, may seem confusing, but these are merely directions for the electrical contractor showing the number of fixtures or outlets to be hooked onto a circuit and the number of circuits radiating from the lighting panel. The wires are by no means installed in the locations as shown.

The plan indicates numerous circuits, such as 1A-14, 1A-7, 1A-16, and so on. The first number represents the lighting panel from which the circuit originates,

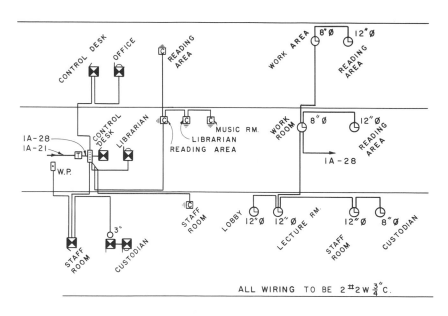

ALL WIRING TO BE 2 #2 W 3/4" C.

LOW TENSION & CLOCKS

Figure 18-2

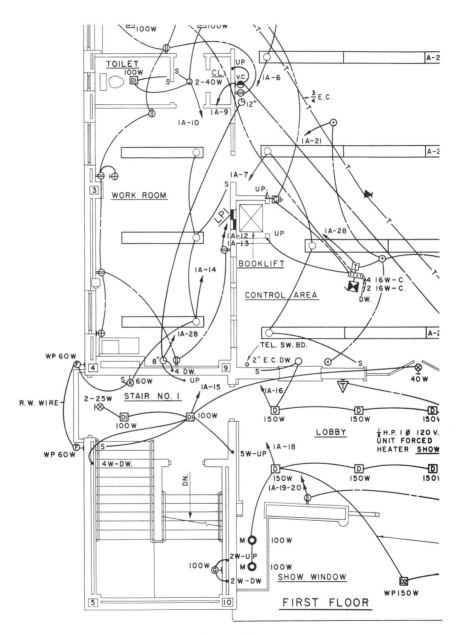

Figure 18-3

the letter indicates the panel section and the second number is the circuit number. Each circuit points to its lighting panel from which it originates. Note the symbol for the lighting panel. The lighting panel, LP1, is located in the workroom wall behind the book lift.

THE TAKE-OFF

Most electrical subcontractors use their own particular method of estimating material quantities and installation costs based on their previous experience with similar jobs. A generally used method, however, is to count the number of outlets, switches, and so on, and multiply the total by a cost per outlet. This includes the wiring required from the lighting panel to the outlets.

Circuits that are housed in conduits will require a somewhat higher cost per outlet.

Other subcontractors, based on their experience, place a cost on a combined group of circuits radiating from a lighting panel. In one area of the country a group of 10 circuits radiating from a lighting panel were quoted by one contractor as costing $405.

Figure 18-4 shows a take-off of outlets and fixtures, together with unit and installation costs, shown on the cellar plan. The plan shows duplex receptacles, and single-pole and three-way switches, as well as outlets for ceiling and wall fixtures. The letters in the ceiling and wall outlets designate the type of fixture, as shown in Fig. 18-5. *Note:* The unit and installation costs given are those of a particular local area and should not be taken as actual costs.

MATERIAL AND INSTALLATION COSTS

Panel board, including circuit breakers, of three-wire 120/240 V and 100 A with 13 circuits costs $526 for material and installation. This includes a 24% contractor's markup for overhead and profit.

Three-wire 120/240 V and 100 A with 16 circuits costs $647 for material and installation. This includes a 24% contractor's markup for overhead and profit.

Three-wire 120/240 V and 100 A with 20 circuits costs $809 for material and installation. This includes a 25% contractor's markup for overhead and profit.

The cost per outlet for circuits without conduits in some areas of the country ranges between $30 and $40, whereas the costs of circuits housed in conduits ranges from $40 to $51.

PARTIAL CELLAR PLAN TAKE-OFF						
Item	Unit Cost	Install. Cost	Unit Plus Install. Cost	Qty.	Total Cost	Overhead and Profit
Outlets						
Duplex receptacles	5.50	13.50	19.00	10	190.00	
Fan outlets	5.50	11.20	16.70	2	33.40	
Single-pole switch	5.50	11.10	16.60	11	182.60	
Three-way switch	7.50	14.20	21.70	2	43.40	
Chimes	32.00	35.00	67.00	1	67.00	
Clocks - 12" Ø	42.00	24.00	66.00	2	132.00	
L.P.C. lighting panel	52.00	78.00	130.00	1	130.00	
					778.40	
Fixtures						
A	22.00	10.00	32.00	5	160.00	
B	20.00	11.00	31.00	1	31.00	
D	23.00	11.05	34.05	6	204.30	
D1	24.00	12.00	36.00	5	180.00	
K	18.00	10.00	28.00	3	84.00	
J	16.00	11.00	27.00	3	81.00	
H	7.00	12.00	19.00	2	38.00	
G	11.00	10.00	21.00	2	42.00	
					820.30	
Fixture Outlets						
A	5.60	13.00	18.60	5	93.00	
B	5.90	11.20	17.10	1	17.10	
D	5.60	12.00	17.60	6	105.60	
D1	5.75	13.00	18.75	5	93.75	
K	5.40	11.10	16.50	3	49.50	
J	5.45	12.00	17.45	3	52.35	
H	5.30	14.00	19.30	2	38.60	
G	5.50	12.10	17.60	2	35.20	
					485.10	
				TOTAL = $ 2083.80		
				ADD 30% FOR O. & P. = $ 2708.94		

Figure 18-4

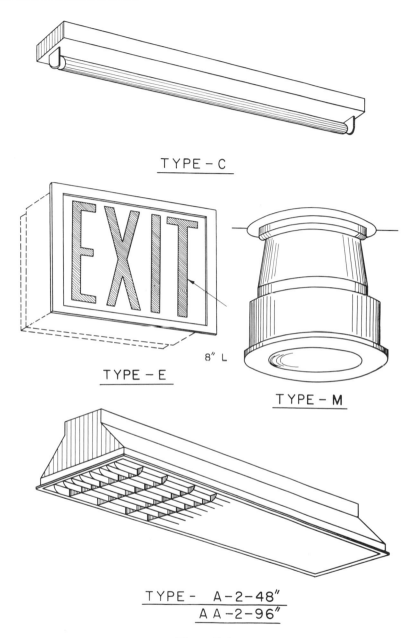

TYPE – C

TYPE – E

8″ L

TYPE – M

TYPE – A-2-48″
AA-2-96″

Figure 18-5

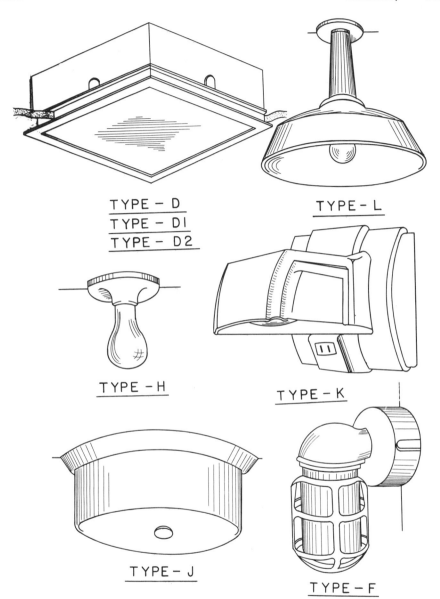

TYPE – D
TYPE – D1
TYPE – D2

TYPE – L

TYPE – H

TYPE – K

TYPE – J

TYPE – F

Figure 18-5 (cont.)

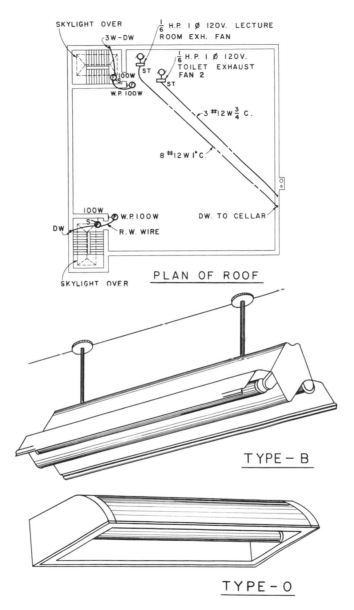

PLAN OF ROOF

TYPE - B

TYPE - 0

Figure 18-5 (cont.)

SELF EXAMINATION

Refer to the second-floor electrical plan (Fig. 18-6) and find the total number of the following:

1. Fluorescent fixtures
2. Duplex receptacles
3. Floor outlets
4. Fan and Switch outlet
5. Vacuum cleaning outlet
6. Single-pole switch
7. Exit outlet
8. Three-way switch
9. Clocks
10. Telephone

Answer Box			
Fluorescent fixtures	16	Single-pole switch	7
Duplex receptacles	7	Three-way switch	1
Floor outlets	3	Exit outlet	1
Fan and switch outlets	4	Clocks	2
Vacuum cleaning outlet	1	Telephone	2

ASSIGNMENT—18

Refer to Fig. 18-6.
1. List the number of circuits.
2. List the number of D1 fixtures.
3. On what circuit are floor outlets shown?
4. List the number of duplex receptacles.
5. List the number of fan and switch outlets.

Refer to Fig. 18-7.
6. How many exit outlets are there, and what is the total wattage?
7. List the number of single-pole switches.
8. How many D1 fixtures are there?
9. The H fixture has how many watts?
10. Where is a K fixture indicated?

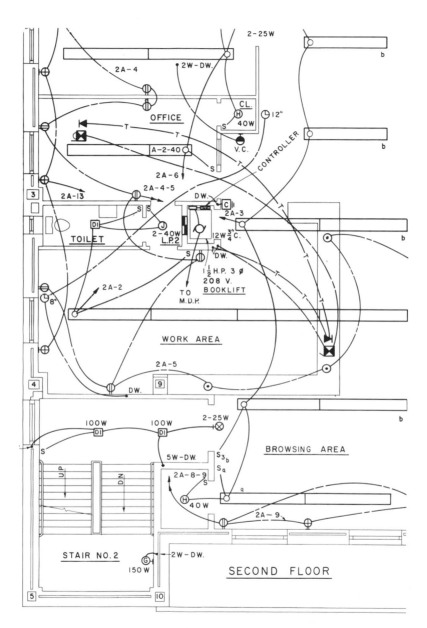

Figure 18-6

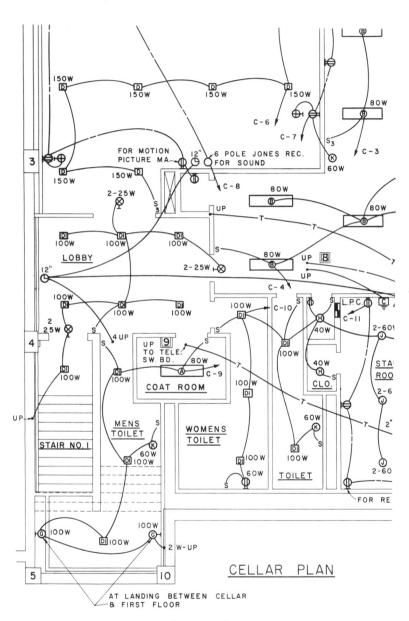

Figure 18-7

SUPPLEMENTARY INFORMATION

WATTAGE CONSUMPTION OF LIGHTS, APPLIANCES, AND OTHER EQUIPMENT

Item	Watts
Television set	280
Radio	35
Air-conditioner	1500
Table lamp	100
Dining room light	150
Vacuum cleaner	720
Sun lamp	280
Room electric heater	1600
Refrigerator	240
Electric range	8000
Automatic toaster	1120
Dishwasher	1750
Clothes washer and dryer	5200
Hot water heater	2500
Cow milker	400

ELECTRICAL SYMBOLS

B 80W	CEILING OUTLET (LETTER DESIGNATES FIXTURE TYPE 100W DES. WA
G 100W	WALL BRACKET OUTLET (DO)
X	EXIT OUTLET
	SINGLE RECEPTACLE 2 POLE
	DUPLEX RECEPTACLE 3 POLE I POLE POLARIZED GROUNDED
V.C.	VACUUM CLEANING RECEPTACLE 3 POLE I POLE GROUNDED 20A
	FLOOR OUTLET
	FAN & SWITCH OUTLET
W.P.	20A - 2P TWISTLOCK RECEPTACLE FOR PROJECTION MACHINE AMPLIFIER & LOUDSPEAKER BACK OF SCREEN
	WEATHERPROOF OUTLET BOX - BRASS
□	RECESSED FIXTURE (JUNCTION BOX
A-40W	FLUORESCENT FIXTURE INDIVIDUAL (LETTER INDICATES TYPE, NUMBER INDICATES WATTAGE OF EACH FIXTURE)
A-40W | A-40W	FLUORESCENT FIXTURE CONTINUOUS
S	SINGLE POLE SWITCH
S_3	THREE WAY SWITCH
S_P	SWITCH & PILOT LIGHT
ST	THERMAL OVERLOAD SINGLE POLE SWITCH
	THERMOSTAT
T	TRANSFORMER
	DESK PUSHBUTTON PAD
C	CHIMES
	TELEPHONE OUTLET
	TERMINAL BOX
3"	MASTER CLOCK, SEMI-FLUSH (CLOCK SEMI-FLUSH TYPE
	FLUSH BELL
	FLUSH BUZZER
	WEATHERPROOF PUSH BUTTON

E L E C T R I C A L   S Y M B O L S (cont.)

⌾	MOTOR
⊏S⊐	STARTER
▰	COMBINATION CONTROLLER & CIRCUIT BREAKER
▬▬	LIGHTING PANEL
▥	MAIN DISTRIBUTION PANEL

———————— INDICATES   CONDUIT INSTALLED   IN FLOOR ABOVE

—·—·—   ''   ''   ''   ''   ''

— T —   ''   ''   ''   ''   ''   ''   FOR TELEPH

Ⓙ   JUNCTION BOX

— – – —   SIGNAL CONDUIT

Ⓟ   PULLBOX

E.C.   EMPTY CONDUIT

⊡⌐   SAFETY SWITCH

————————   CIRCUIT DESIGNATION
  I A2

  FIRST NUMERAL DENOTES LIGHTING PANEL NUMBER
  LETTER INDICATES PANEL SECTION
  SECOND NUMERAL INDICATES CIRCUIT NUMBER

Fluorescent-Tube Lighting

A fluorescent tube gives about twice as much light for the same number of watts as an incandescent bulb and lasts about $2\frac{1}{2}$ times as long.   A two-tube fluorescent fixture rated at 80 W gives as much light as a 150-W incandescent light bulb.

FLUORESCENT BULBS

Wattage	Length (in.)	Tube diameter (in.)
14	15	$1\frac{1}{2}$
15	18	$1\frac{1}{2}$
15	18	$1\frac{1}{2}$
20	24	$1\frac{1}{2}$
30	36	$1\frac{1}{2}$
40	48	$1\frac{1}{2}$
	Circular:	1
20–40	$8\frac{1}{2}$, 12, 16	
	diameter	

19

MATHEMATICS FOR THE ESTIMATOR

INTRODUCTION

This unit deals with the basic mathematics and manipulations used by the estimator. Although many of these manipulations are not unfamiliar, it is well to review them here since a decimal point in the wrong place can mean the loss of a job or the loss of large sums of money.

TECHNICAL INFORMATION

Addition—Subtraction—Multiplication

In addition and subtraction remember that all decimal points are placed directly under each other.   Look at the following examples:

Add 6.14 + 0.03 + 13.7:   Subtract 367.40 from 9125.25:

$$
\begin{array}{r}
6.14 \\
0.03 \\
+\ 13.70 \\
\hline
19.87
\end{array}
\qquad
\begin{array}{r}
9125.25 \\
-\ 367.40 \\
\hline
8757.85
\end{array}
$$

In multiplication, point off in the answer as many places from the right as there are digits or decimal places to the right of the multiplier and the multiplicand. Look at the following multiplication problem:

1.75 (one-point-seven-five) is multiplied by 0.25 (point-two-five)

$$
\begin{array}{r}
1.75 \\
\times\ 0.25 \\
\hline
875 \\
350 \\
\hline
0.4375
\end{array}
$$

Both the multiplicand and the multiplier have two decimal places to the right of the decimal point, or a total of four decimal places.   In the answer, therefore, four decimal places are pointed off.

Taking a Percentage of a Number

Other important manipulations required by the estimator are percentages of numbers. Suppose that we take 30% of 125 (one-twenty-five).   This is found by multiplying 125 by 0.30:

$$
\begin{array}{r}
125 \\
\times\ 0.30 \\
\hline
37.50
\end{array}
$$

Adding a Percentage to a Number

Suppose that we add 16% to 196.   First multiply 196 by 0.16, and then add the answer to 196:

$$
\begin{array}{r}
196 \\
\times\ 0.16 \\
\hline
1176 \\
196 \\
\hline
31.36
\end{array}
\qquad
\begin{array}{r}
196 \\
+\ \ 31.36 \\
\hline
227.36
\end{array}
$$

A simpler method, or shortcut, is to multiply 196 by 1.16:

$$
\begin{array}{r}
196 \\
\times\ 1.16 \\
\hline
1176 \\
196 \\
196 \\
\hline
227.36
\end{array}
$$

Division of Numbers

In division, the decimal point is always moved to the right and placed after the last digit in the divisor. At the same time an equal number of places is counted off to the right in the dividend.

For example, in the problem 36 ÷ 1.2, the decimal point of the divisor 1.2 is moved one place to the right so that the number reads 12. At the same time, one decimal point in the dividend 36 is moved to the right so that the number reads 360. Directly above the decimal point in the dividend is placed the decimal point of the quotient or answer:

$$\text{Divisor } 1.2 \,\overline{)36} \text{ divided change to } 12\overline{)360}$$
$$\begin{array}{r} 30 \text{ quotient} \\ 36 \\ \hline 0 \end{array}$$

Examples in placing the decimal point:

$$37.1 \,\overline{)16.75} \quad \text{change to} \quad 371 \,\overline{)167.5}$$

$$0.765 \,\overline{)2.043} \quad \text{change to} \quad 765 \,\overline{)2043.}$$

$$6.3 \,\overline{)7.65} \quad \text{change to} \quad 63 \,\overline{)76.5}$$

$$3.2 \,\overline{)587.3} \quad \text{change to} \quad 32 \,\overline{)5873.}$$

How are inches changed to the decimal part of a foot? Figure 19-1 gives the decimal parts of a foot for inches from 1 to 11, including $\frac{1}{4}$ and $\frac{1}{2}$ in. One inch, for example, is read 0.08 (pronounced point-zero-eight) of a foot. Two inches is 0.17 (point-one-seven) of a foot, 3 in. is 0.25 (point-two-five) of a foot, and so on.

How are inches converted into the decimal parts of a foot? One inch is $\frac{1}{12}$ of a foot; 2 in. are $\frac{2}{12}$ of a foot; 3 in. are $\frac{3}{12}$ of a foot, and so on. Divide the inches

INCHES TO THE DECIMAL PART OF A FOOT												
$\frac{1}{4}$"	$\frac{1}{2}$"	1"	2"	3"	4"	5"	6"	7"	8"	9"	10"	11"
0.02	0.04	0.08	0.17	0.25	0.33	0.42	0.5	0.58	0.67	0.75	0.83	0.92

Figure 19-1

by 12 to arrive at the decimal part of a foot: for example, 4 in. is 4 in. ÷ 12 in., or

$$12 \overline{)\begin{array}{l} 0.33 \text{ ft} \\ 4.0 \\ \underline{36} \\ 40 \\ \underline{36} \end{array}}$$

The decimal parts of a foot from 1 to 11 in., including $\frac{1}{4}$ and $\frac{1}{2}$ in., should be memorized. When taking-off dimensions from the plans, the estimator will automatically change the inches to the decimal part of a foot and enter the decimal figure on the take-off form.

Reading Decimal Dimensions from the Plans

On the plan in Fig. 19-2, the 40′6″ dimension will read 40.5 (forty-point-five) ft. The 20′9″ dimension will read 20.75 (twenty-point-seven-five) ft. It is well to practice reading all the dimensions in this plan in their decimal form in order to gain the familiarity required by the estimator.

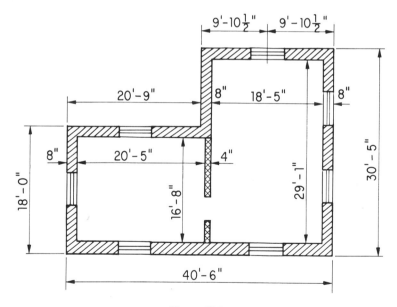

Figure 19-2

SELF EXAMINATION

Now you have the opportunity to find out how well you understand the material thus far. Work out the problems of the self examination and check your answers with the key answers in the Answer Box.

1. Add 3.17 + 81.125 + 0.092.
2. Subtract 675.75 − 18.029.
3. Multiply 13.07 × 4.152.
4. Take 37% of 1165.
5. Add 33% to 1250.
6. Divide 125.12 by 9.2.
7. Change 5 in. to the decimal part of a foot.
8. Write the following dimensions in decimal form: 19′8″, 36′9″, 47′7′, 12′0½″, 6′1¼″, and 16′3″.

Answer Box
1. 84.387, or 84.39
2. 657.721, or 657.72
3. 54.26664, or 54.27
4. 431.05
5. 1662.5
6. 13.6
7. 0.416 or 0.42 ft
8. 19.67 ft
36.75 ft
4.58 ft
12.04 ft
6.10 ft
16.25 ft

Finding Square Feet

The rectangle in Fig. 19-3 measures 10′6″ × 16′4″.   To find the area or the number of square feet, multiply the length by the width of the rectangle.

The thing to watch here is to change the inches to the decimal part of a foot, so that 16′4″ will read 16.33 ft and the 10′6″ dimension will read 10.5 ft.   By multiplying 16.33 ft by 10.5 ft, the answer is 171.465, or 171.47 sq ft.   In estimating, two figures to the right of the decimal point are sufficient.   If the third figure is five or more, the second figure is increased by one.

The Area of a Triangle

The area in Fig. 19-4 is found by multiplying the height by the base of the triangle and dividing the product in half.
  For example, 5′9″ × 3′3″ is changed to

$$5.75 \text{ ft} \times 3.25 \text{ ft} = 18.6875 \text{ sq ft}$$

$$18.6875 \text{ sq ft} \div 2 = 9.34375, \text{ or } 9.34, \text{ sq ft}$$

The last three figures may be dropped.

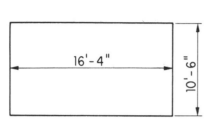

Figure 19-3   **Figure 19-4**

The Area of a Circle

The area in Fig. 19-5 is found by multiplying π (pronounced pi or pie), a Greek letter representing the number 3.14, by the radius squared.
  For example, the radius of the 10-in.-diameter circle is 5 in.   (The radius is always one-half of the diameter.)   The term "radius squared" means to multiply the radius by itself, such as 5 ft × 5 ft = 25 sq ft (the radius squared); then multiply 25 ft × 3.14 = 78.50 sq ft, the area of the circle.

Area of a Trapezoid (Fig. 19-6)

Add the two bases of the trapezoid, multiply the sum by the height, and then divide by two.   The formula is

$$A = \frac{H(B_1 + B_2)}{2}$$

where A = area, B = height of trapezoid, B_1 = base 1 (bottom), and B_2 = base 2 (top).
  Note: The parentheses in the formula indicate multiplication.   The value for H is multiplied by the sum of the values for B_1 and B_2.   The 2 under the line means to divide by 2.
  For example, find the area of a trapezoid having a bottom base of 8′0″, a top

base of 4'0", and a height of 3'6". Then

$$A = \frac{3.5(8 + 4)}{2} = \frac{3.5(12)}{2} = \frac{42}{2} = 21 \text{ sq ft}$$

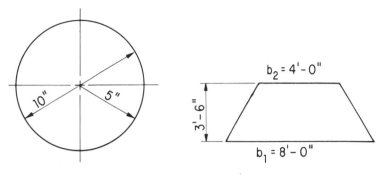

Figure 19-5   Figure 19-6

Circumference of a Circle

Multiply pi (or 3.14) by the diameter of the circle. (*Note:* The circumference is the distance around the circle.) The formula is

$$C = \pi \times D$$

where C represents circumference, π is 3.14, and D is the diameter.

For example, find the circumference of a circle whose diameter is 5 in. Then,

$$C = \pi \times D$$
$$C = 3.14 \times 5 \text{ in.} = 15.70 \text{ in.}$$

Length of a Curved Wall (Fig. 19-7)

The centerline radius of the curved wall is 20'6". The most correct length of the wall is found by using the centerline radius of the circle. Then

$$C = \pi \times D$$
$$= 3.14 \times 41 \text{ ft} = 128.74 \text{ ft} \div 2 \text{ (half-circle)}$$
$$= 64.37 \text{ ft}$$

Finding the Square Yards (Fig. 19-8)

To find the square yards of a surface such as a wall or floor, find the square feet and divide by 9. (There are 9 sq ft in 1 sq yd)

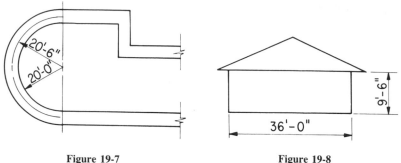

Figure 19-7   **Figure 19-8**

For example, the exterior wall surface of the house in Fig. 19-7 measures 36′0″ by 9′6″. Then,

$$36 \text{ ft } \times 9.5 \text{ ft } = 342 \text{ sq ft}$$

$$342 \text{ sq ft } \div 9 = 38 \text{ sq yd}$$

Finding the Cubic Feet (Fig. 19-9)

To find the cubic content of the wall in fig. 19-9, which measures 8 in. thick by 9 ft high by 16 ft long, multiply the width by the height by the length.   Remember to change the inches to the decimal part of a foot before multiplying.
    For example, 8 in. is 0.67 ft.   Then,

$$0.67 \text{ ft } \times 9 \text{ ft } \times 16 \text{ ft } = 96.48 \text{ cu ft}$$

Finding Cubic Yards

There are 27 cu ft in 1 cu yd.   Therefore, to change cubic feet into cubic yards, divide by 27.   For example, assume a footing and foundation wall as in Fig. 19-10.   Find the total cubic yards of concrete required.
    The footing is

$$2 \text{ ft } \times 1 \text{ ft } \times 10 \text{ ft } = \quad 20 \text{ cu ft}$$

The wall is

$$1 \text{ ft } \times 8 \text{ ft } \times 10 \text{ ft } = \quad \underline{80 \text{ cu ft}}$$

$$100 \text{ cu ft}$$

Therefore,

$$100 \text{ cu ft } \div 27 = 3.66 \text{ cu yd}$$

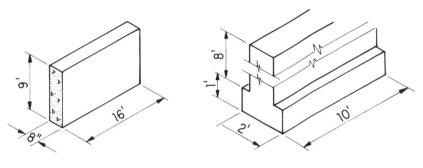

Figure 19-9   Figure 19-10

Finding the Cubic Feet of a Cylinder

First find the area of the circular face of the cylinder in Fig. 19-11. Multiply this by the height of the cylinder to get the cubic feet.

For example, find the cubic feet of a cylinder whose diameter is 4'0″ and which is 8'0″ high.

$$3.14 \times 2 \text{ ft} \times 2 \text{ ft} = 12.56 \text{ sq ft} \times 8 \text{ ft} = 100.48 \text{ cu ft}$$

Board Measure—What Does It Mean?

A board foot of lumber is equal to 144 cu in. of wood. A board that measures 12 in. × 12 in. × 1 in. in thickness equals 144 cu in., which is 1 B.F.

A simple rule in finding the board feet of any piece of lumber is to multiply the cross-sectional dimension of the lumber in inches by the length of the piece in feet and to divide by 12.

For example, applying the rule to Fig. 19-12, we proceed as follows:

$$\frac{1 \text{ in.} \times \cancel{12} \text{ in.} \times 1 \text{ ft}}{\cancel{12}} = 1 \text{ B.F.}$$

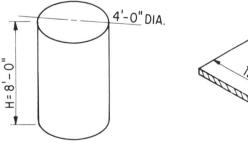

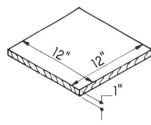

Figure 19-11   Figure 19-12

Similarly, the board feet of a 2 in. × 4 in. piece of studding 8 ft long can be found as follows:

$$\frac{2 \text{ in.} \times 4 \text{ in.} \times 8 \text{ ft}}{\underset{3}{12}} = \frac{16}{3} = 5\tfrac{1}{3} \text{ B.F.}$$

SELF EXAMINATION

1. Find the area of a rectangle that measures 22'3" × 9'6".
2. Find the area of a triangle whose base is 3'3" and whose height is 5'9".
3. Find the area of a circle having a 20-ft diameter.
4. Find the area of a trapezoid, the bases of which are 9 ft and 3 ft. The height is 4 ft.
5. The diameter of a circle is 11 ft. What is its circumference?
6. Find the square yards of the walls and ceiling of a room that measures 12 ft × 16 ft. The ceiling height is 8 ft.
7. Find the cubic feet of the room in Problem 6.
8. A truck is loaded with 162 cu ft of sand. Find the cubic yards.
9. Find the cubic feet in a cylinder that has a 5-ft diameter and is 10 ft high.
10. 60 pieces of 2 × 4's, 8 ft in length, contain how many board feet?

Answer Box
1. 211.38 sq ft
2. 9.34 sq ft
3. 314 sq ft
4. 24 sq ft
5. 34.54 in.
6. 71.33 sq yd
7. 1536 cu ft
8. 6 cu yd
9. 196.30 cu ft
10. 320 B.F.

ASSIGNMENT—19

1. A builder paid $36,000 for six lots. On each lot he built houses that cost per house: $12,000 for materials, $14,000 for labor, and 10% of total cost per house and lot for surveys, legal fees, brokers fees, and so on. Each house sold for $42,500. Determine the dollar and percentage profit per house.

2. A builder purchased lots as shown in Fig. 19-13.
   (a) Determine the number of acres purchased (1 acre = 43,560 sq ft).
   (b) If the builder paid $0.50/sq ft for lots, what is the cost of lot 101? How much should the builder sell it for to make a 15% profit?

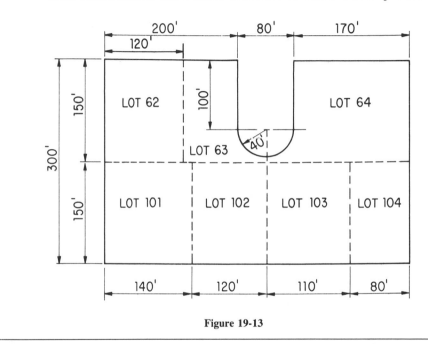

Figure 19-13

SUPPLEMENTARY INFORMATION

Areas of Plane Figures

Square.   Square the length of one side.

$$A = L^2$$

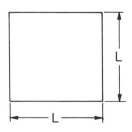

Rectangle.   Multiply length by width.

$$A = LW$$

Triangle.   Multiply half the altitude by the length of the base.

$$A = \frac{AB}{2}$$

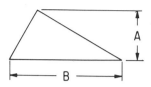

Circle.   Square the diameter, and multiply by 0.7854.

$$A = 0.7854D^2$$

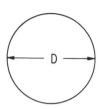

Ellipse.   Multiply the minor axis by the major axis by 0.7854.

$$A = 0.7854Dd$$

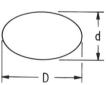

Hexagon.   Square the short diameter and multiply by 0.886, or square the long diameter and multiply by 0.6495.

$$A = 0.866d^2$$

or

$$A = 0.6495D^2$$

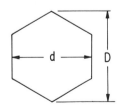

Mensuration

Octagon.   Square the short diameter and multiply by 0.828, or square the long diameter and multiply by 0.707.

$$A = 0.828d^2$$

$$A = 0.707D^2$$

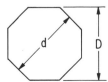

Parallelogram.   Multiply length by perpendicular height.

$$A = LH$$

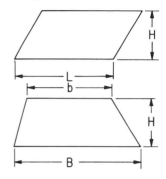

Trapezoid.   Multiply height by half the sum of the top and bottom bases.

$$A = \frac{H(b + B)}{2}$$

Volumes of Typical Solids

Any prism or cylinder, right or oblique, rectangular or not:

> volume = area of base × altitude
>
> altitude = distance between parallel bases, measured
>
> perpendicular to the bases

When bases are not parallel:

> altitude = perpendicular distance from one base to the
>
> center of the other

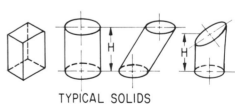

TYPICAL SOLIDS

REVIEW QUESTIONS

1. Read the following dimensions in decimal parts of a foot: 12′9″, 3′6″; 8′7″; 3′4″, 2′11″; 12′5″.
2. Write the formula for finding the area of a circle.
3. Write the formula for finding the circumference of a circle.
4. How do you find the area of a trapezoid?
5. Give the dimensions of a piece of lumber that equals 1 B.F.

20

REVIEW PROBLEMS
OR TESTS

EXCAVATION (Fig. 20-1)

Take-off the following:

1. Remove topsoil 6 in. deep, 10'0" beyond exterior walls.
2. General excavation 2'0" beyond exterior walls.
3. Hand excavation for footing trenches.

Item	Unit	Length	Width	Depth	Cubic Feet	Cubic Yards
Topsoil						
General excavation						
Footing trench						

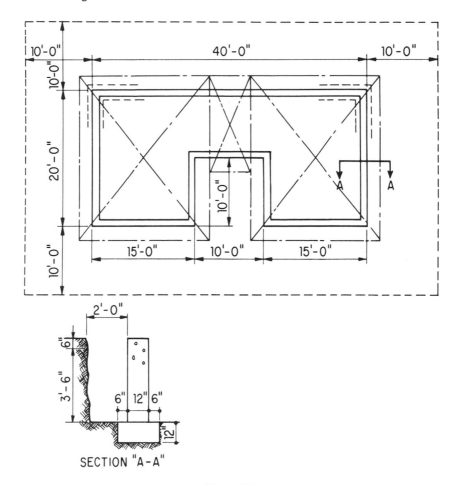

Figure 20-1

CONCRETE FOOTINGS AND FOUNDATION WALLS   (Fig. 20-2)

Take-off the following:

1. Concrete footings.
2. Concrete foundation walls.
3. 4-in. slab.
4. 6-in. cinder fill.

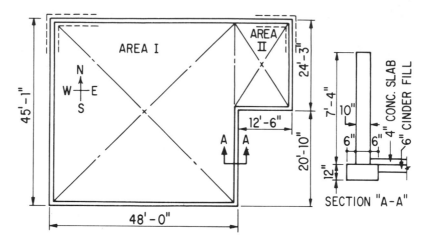

Figure 20-2

Item	Unit	Length	Width	Depth	Cubic feet	Cubic yards

Concrete footings
  North and south
  East and west						

Foundation walls
  North and south
  East and west						

4-in. concrete
   slabs
  Area I
  Area II						

6-in. cinder fill
  Area I
  Area II						

CONCRETE WALLS, SLABS, CEMENT FINISH, AND TILE   (Fig. 20-3)

Take-off the following:

1. Concrete foundation wall.
2. Concrete footing (12 in. × 24 in.).
3. 4-in. concrete slab

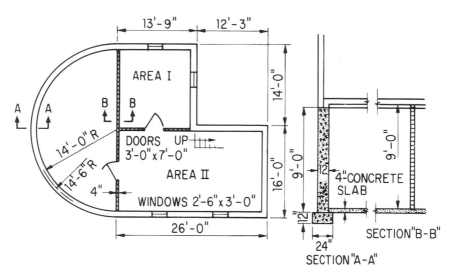

Figure 20-3

4. Glazed tile partition blocks (4 in. × 12 in.).   Allow 3% for waste (nominal dimension).

Item	Unit	Length	Width	Depth	Cubic feet	Cubic feet out	Cubic yards
Foundation walls Curved Straight Windows							
Concrete footings Curved area Straight							
4-in. concrete slab, Curved area Area I Area II			(square feet)				
Glazed tile Doors					(square feet)		(number of blocks)

CONCRETE BLOCKS   (Fig. 20-4)

Take-off the following:

1. Concrete block 8 in. × 12 in. × 16 in.
2. Concrete block 8 in. × 8 in. × 16 in., 112.5 blocks/100 sq ft.   Add 2% for waste.

Item	Unit	Length	Height	Square feet	Hundredths of square feet	Number of blocks
Building 1						
Building 2 Building 3 Building 4						

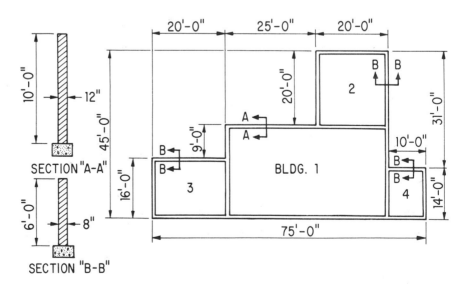

Figure 20-4

BRICK (Fig. 20-5)

Take-off the following:

1. Common standard brick, 8 in. $\times$ $2\frac{1}{4}$ in. $\times$ $3\frac{3}{4}$ in. Add 2% for waste. $\frac{1}{2}$-in. mortar joint.
2. Quantity of mortar.

Item	Unit	Length	Height	Square feet	Square feet Out	Bricks per square foot	Actual brick	Total brick
Brick								
Windows (out)								
Door (out)								
Mortar								

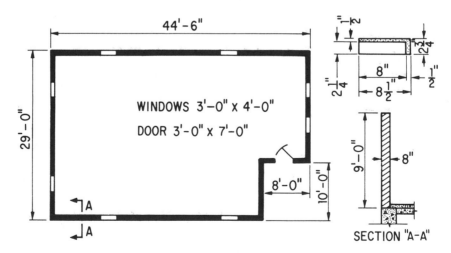

Figure 20-5

LUMBER   (Fig. 20-6)

Take-off the following:

1. Sill plate.
2. Joist and header.
3. Sole plate.
4. Top plate.
5. Studs.

Find the board feet of all items.
Cost = $0.15/B.F.
Find the costs.

Item	Unit	Length feet	Total length	Board feet	Cost per board foot	Total cost
2 × 6 Sill						
2 × 8 Joists 2 × 8 Header						
Sole plate						
Top plate						
Studs						

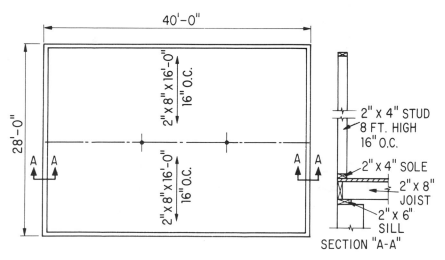

Figure 20-6

WIRE LATH AND PLASTER   (Fig. 20-7)

Take-off the following:

1. Metal lath for rooms and ceilings marked 1, 2, and 3.   Use 10-sheet bundles; each sheet is 27 in. × 96 in. (20 sq yd per bundle).
2. Plaster on walls and ceilings of rooms 1, 2, and 3.   Material quantities per 100 sq yd of plaster:
   Scratch coat—10 (100-lb) bags gypsum plaster, 1 cu yd sand.
   Brown coat—7 (100-lb) bags gypsum plaster, 21 cu ft sand.
   Finish coat—7 (50-lb) bags lime and 150 lb gauging plaster.

METAL LATH

Item	Unit	Length	Width	Height	Square feet	Square yards	Bundles
Walls							
Ceilings							

PLASTER

Ceiling square feet	Scratch	Bags plaster Cubic yards sand
Walls square feet (less 7% for openings)	Brown	Bags plaster Cubic feet sand
Total square feet	Finish	Bags lime Pounds gauging Plaster
Total square yards		

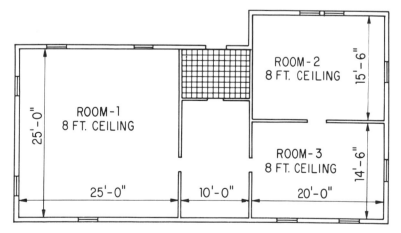

Figure 20-7

PAINTING   (Fig. 20-8)

Take-off the following:

1. Square feet of striated wall shingles to be painted (first coat) with acrylic base paint.   Large window is 50′0″ × 8′0″.
2. Gallons of paint required.
3. Labor time.

Item	Unit	Length	Height	Square feet	Gallons	Hours
Walls						
Gables						

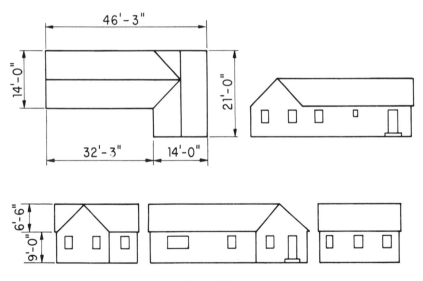

Figure 20-8

21

PLANS FOR PRACTICE ESTIMATING

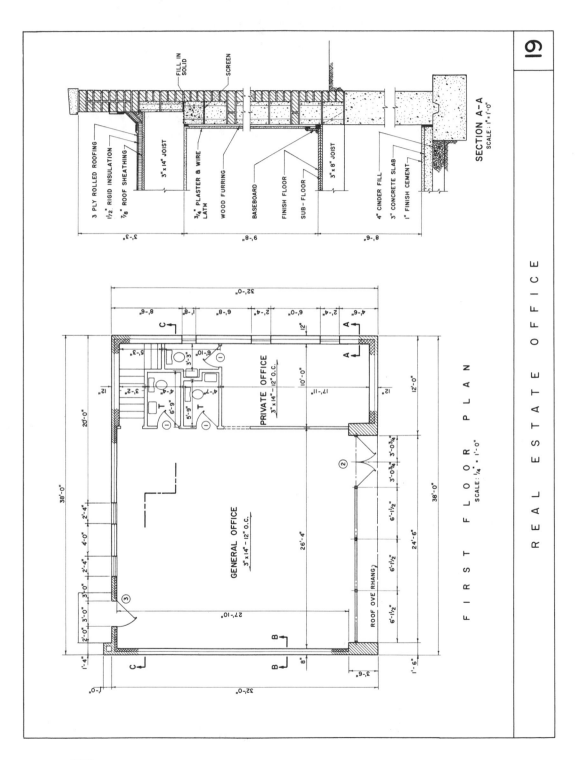

SECTION A-A
SCALE: 1" = 1'-0"

3 PLY ROLLED ROOFING
1½" RIGID INSULATION
⅞" ROOF SHEATHING
3" x 14" JOIST
¾" PLASTER & WIRE LATH
WOOD FURRING
BASEBOARD
FINISH FLOOR
SUB - FLOOR
3" x 8" JOIST
FILL IN SOLID
SCREEN
4" CINDER FILL
3" CONCRETE SLAB
1" FINISH CEMENT

3'-3"
9'-8"
8'-6"

FIRST FLOOR PLAN
SCALE: ¼" = 1'-0"

PRIVATE OFFICE
3" x 14" — 12" O.C.

GENERAL OFFICE
3" x 14" — 12" O.C.

ROOF OVE RHANG

32'-0"
38'-0"

REAL ESTATE OFFICE

19

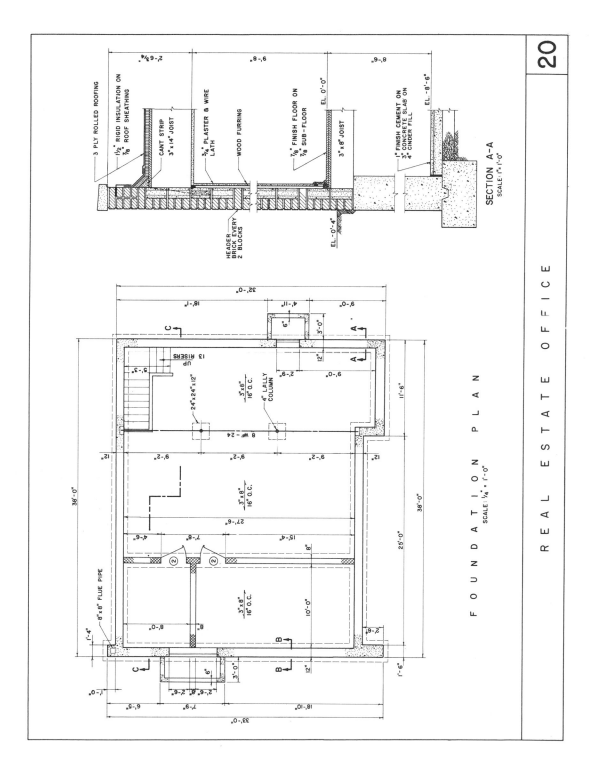

SECTION A-A
SCALE: 1" = 1'-0"

3 PLY ROLLED ROOFING

1½" RIGID INSULATION ON
⅞ ROOF SHEATHING

CANT STRIP

3" x 14" JOIST

¾" PLASTER & WIRE
LATH

WOOD FURRING

HEADER
BRICK EVERY
2 BLOCKS

⅞" FINISH FLOOR ON
⅞ SUB-FLOOR

3" x 8" JOIST

EL. 0'-0"

1" FINISH CEMENT ON
3" CONCRETE SLAB ON
4" CINDER FILL

EL. -0'-4"

EL. -8'-6"

2'-6¾"

9'-8"

8'-6"

FOUNDATION PLAN
SCALE: ¼" = 1'-0"

32'-0"

18'-1"

9'-0"

4'-11"

6"

3'-0"

12"

UP
13 RISERS

5'-3"

2'-6"

9'-0"

24"x24"x12"

3"x6"
16" O.C.

4" LALLY COLUMN

8 WF-24

C

38'-0"

12"

9'-2"

9'-2"

9'-2"

12"

11'-6"

38'-0"

3"x8"
16" O.C.

27'-6"

4'-6"

15'-4"

7'-8"

8"

25'-0"

2
2

3"x8"
16" O.C.

10'-0"

8'-0"

8"

8"

2'-6"

8"x8" FLUE PIPE

1'-4"

1'-0"

6'-5"

7'-9"

18'-10"

33'-0"

6"

2'-6"

3'-0"

B

12"

1'-6"

A

A

B

REAL ESTATE OFFICE

20

283

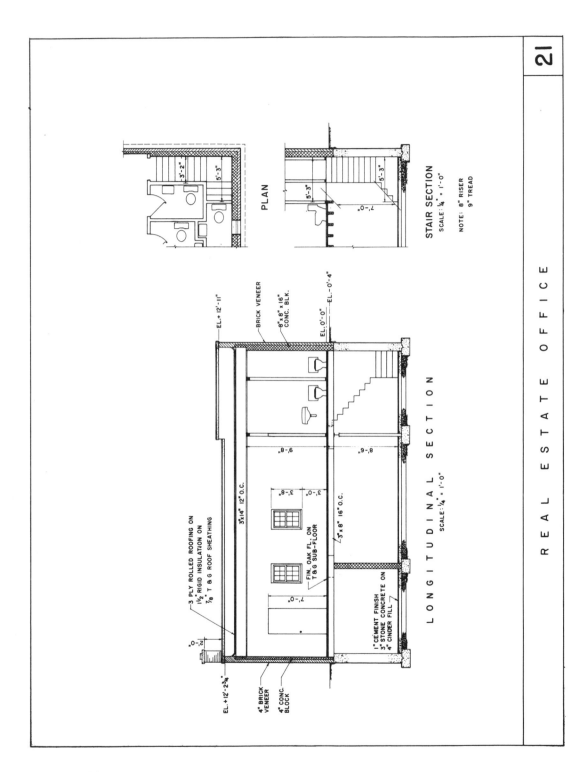

PLAN

STAIR SECTION
SCALE: ¼" = 1'-0"

NOTE: 8" RISER
9" TREAD

LONGITUDINAL SECTION
SCALE: ¼" = 1'-0"

3 PLY ROLLED ROOFING ON
1½" RIGID INSULATION ON
⅞" T & G ROOF SHEATHING

EL.+ 12'-11"

BRICK VENEER

8"x 8" x 16"
CONC. BLK.

EL.+ 0'-0"
EL.-0'-4"

3 x 14" 12" O.C.

FIN. OAK FL. ON
T & G SUB-FLOOR

3" x 8" 16" O.C.

1" CEMENT FINISH
3" STONE CONCRETE ON
4" CINDER FILL

EL.+ 12'-2¾"

4" BRICK
VENEER

4" CONC.
BLOCK

R E A L   E S T A T E   O F F I C E

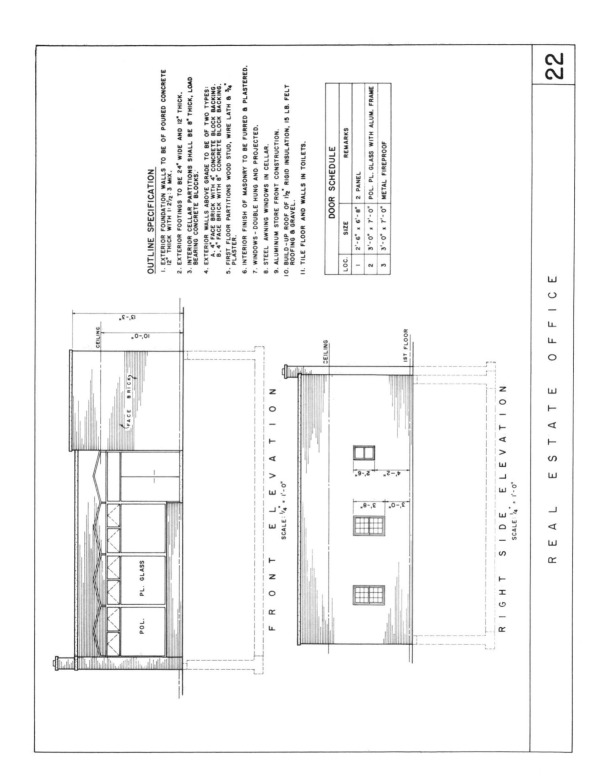

OUTLINE SPECIFICATION

1. EXTERIOR FOUNDATION WALLS TO BE OF POURED CONCRETE 12" THICK WITH 1: 2½: 3 MIX.

2. EXTERIOR FOOTINGS TO BE 24" WIDE AND 12" THICK.

3. INTERIOR CELLAR PARTITIONS SHALL BE 8" THICK, LOAD BEARING CONCRETE BLOCKS.

4. EXTERIOR WALLS ABOVE GRADE TO BE OF TWO TYPES:
   A. 4" FACE BRICK WITH 4" CONCRETE BLOCK BACKING.
   B. 4" FACE BRICK WITH 8" CONCRETE BLOCK BACKING.

5. FIRST FLOOR PARTITIONS WOOD STUD, WIRE LATH & ¾" PLASTER.

6. INTERIOR FINISH OF MASONRY TO BE FURRED & PLASTERED.

7. WINDOWS — DOUBLE HUNG AND PROJECTED.

8. STEEL AWNING WINDOWS IN CELLAR.

9. ALUMINUM STORE FRONT CONSTRUCTION.

10. BUILD-UP ROOF OF 1½" RIGID INSULATION, 15 LB. FELT ROOFING & GRAVEL.

11. TILE FLOOR AND WALLS IN TOILETS.

DOOR SCHEDULE

LOC.	SIZE	REMARKS
1	2'-6" x 6'-8"	2 PANEL
2	3'-0" x 7'-0"	POL. PL. GLASS WITH ALUM. FRAME
3	3'-0" x 7'-0"	METAL FIREPROOF

FRONT ELEVATION
SCALE: ¼" = 1'-0"

RIGHT SIDE ELEVATION
SCALE ¼" = 1'-0"

REAL ESTATE OFFICE

22

FREEHAND SKETCHING

INTRODUCTION

Freehand sketching demands two basic considerations: the use of the drawing pencils and the ability to sketch to proportion.

The first is of less importance, but essential in using the correct grade of lead pencil. The beginner will invariably reach for a drafting pencil, which is far too hard to be of any use in freehand sketching.

Freehand sketching demands soft-lead pencils such as HB, B, and 2B, sharpened to conical or wedge-shaped points as the sketch demands.

Sketching to proportion is perhaps of greatest importance, since a sketch that is out of proportion is hardly recognizable and has little value. To sketch to proportion, the sketcher must have a knowledge of the relative dimensions of the object that he or she is about to sketch.

For example, let us consider an ordinary 2 in. × 4 in. stud (Fig. 22-1) viewed from its end. The 4 in. is double the 2 in. and should appear that way.

Now suppose that we draw the line AB, any length, representing the width of the stud. To sketch the thickness of the stud, visualize half the stud width and sketch the distance for the thickness of the stud.

Once you have established a dimension—be it width, length, or height— relate these dimensions to surrounding dimensions in order to stay within proportion.

In sketching plans, elevations, sections, and other construction details, the

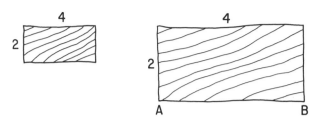

Figure 22-1

sketcher is concerned with three distinct types of object surfaces:

1. *Regular plane surfaces:* The regular plane surface object (Fig. 22-2) is shown in three-dimensional pictorial form, and in two-dimensional projection form. The principal or front view of the three-view drawing is the most descriptive shape of the object. The top view is projected or placed directly above the principal view, while the side view is placed to the side of the principal view. This is the conventional way to show view drawings.

   The principal view can very well be compared to the front elevation of the house, the side view to the side elevation, and the top view to the roof plan of the house. The plan of the house, however, is a section or horizontal cut through the elevation at the midheight of doors and windows, viewed from the top (Fig. 22-3).

2. *Inclined plane surfaces:* Objects that contain neither horizontal nor vertical surfaces are known as inclined plane surfaces. The shaded portion (Fig. 22-4) is the inclined plane. On the three-view drawing, the inclined plane is also represented by shaded portions. In the side view the inclined plane is represented by the dashed line. Without the dashed line the object could not be fully understood.

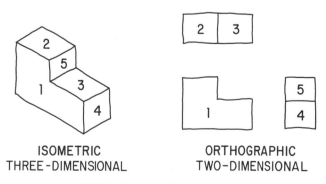

ISOMETRIC
THREE-DIMENSIONAL

ORTHOGRAPHIC
TWO-DIMENSIONAL

REGULAR SURFACE OBJECT

Figure 22-2

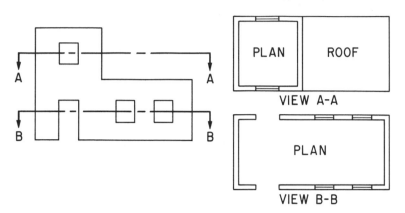

Figure 22-3

3. *Curved surfaces:* Lines on both plane and inclined-plane surfaces are caused by the intersection of two surfaces, causing an edge. The outline of a tabletop is caused by the intersection with the sides of the table. On curved surfaces, however, object lines are not always edge lines (Fig. 22-5) but are sometimes known as tangent lines. The top horizontal line of the front view is a tangent line. It is not caused by the intersection of two surfaces but merely represents the topmost part of the curved object.

Learning to Sketch

Learning to sketch is to begin with simple things, such as lines, rectangles, circles, and other shapes, followed by oblique and isometric pictorial sketches.

Most of the freehand line work is done with HB or B pencils; sometimes softer pencils such as 2B and 4B are used for darker outlines.

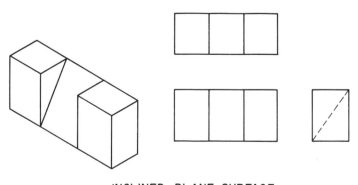

INCLINED PLANE SURFACE

Figure 22-4

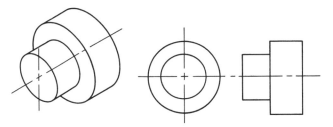

CURVED SURFACE OBJECT

Figure 22-5

Line Exercises

Start sketching practice with simple lines, by drawing freehand, horizontal, vertical, slant, and curved lines (Fig. 22-6). Short lines are made with finger and wrist movement, while longer lines result from a series of short lines. Usually, a point can be made denoting the start of the line, and a point is spotted where the line is to terminate (Fig. 22-7).

A line can quickly be divided into equal parts, by eye, by spotting the center of the line, resulting in two equal halves (Fig. 22-7). Each half can again be subdivided into quarters, and so on. It is important here to visualize equal halves by eye. A little practice will bring results.

Sketching the Square

Lay off two centerlines, vertical and horizontal (Fig. 22-8). Equidistant from the center, mark off four points on the centerlines. Draw horizontal lines and then vertical lines through the points to complete the square.

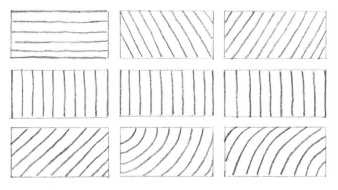

Figure 22-6

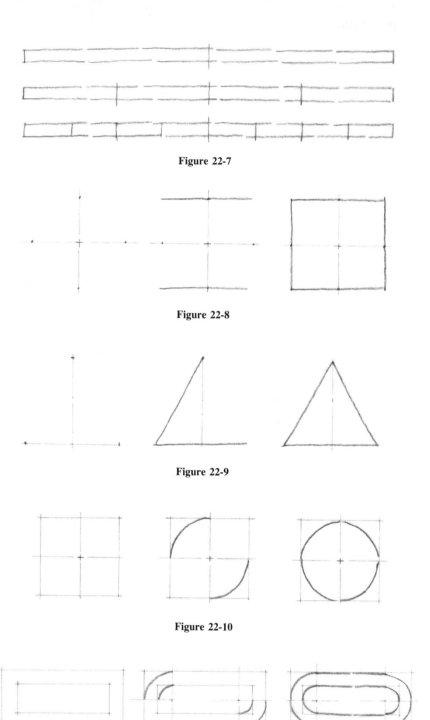

Figure 22-7

Figure 22-8

Figure 22-9

Figure 22-10

Figure 22-11

Sketching the Triangle

Draw a vertical and a baseline (Fig. 22-9). Lay off points on the baseline equidistant from the center of the base. Spot a point on the vertical line which is to be the apex of the triangle. Draw a line from the apex to the left point on the base. Complete the triangle by drawing a line from the apex to the right point on the base.

Sketching the Circle

Begin the circle by drawing a horizontal and vertical centerline (Fig. 22-10). Lay off equidistant from the center points on both centerlines. Draw a square through the four points. Scribe an arc, freehand, in one quadrant and the opposite quadrant. Next, complete the circle by scribing the arcs for the other two quadrants.

Sketching a Chain Link

Draw the rectangle 1, 2, 3, 4 and the inside rectangle a, b, c, d. Divide rectangle into thirds, by eye (Fig. 22-11). Draw the upper left quadrant curve and the lower right quadrant curve. Do the same for the inner quadrants. Next, do all opposite quadrants.

Sketching the Floor Plan

Lightly, at first, draw the exterior wall lines with finger–wrist-movement lines. Draw a second line parallel to the wall outline. Indicate doors and windows. With B or HB pencil, fill in exterior walls but not windows and door openings (Fig. 22-12).

Oblique and Isometric Sketching

Although an object can be sketched in many ways and viewed from any position, the use of either oblique or isometric drawn objects are predominantly used. Figures 22-13 to 22-15 show steps in sketching such pictorial objects in both oblique and isometric views, having regular, inclined-plane, and curved surfaces.

For example, in Fig. 22-13 the oblique drawing is started by sketching the front view of the object as shown in step 1. Lines are then drawn at an angle of about 30° or 45°, by eye, in step 2. The object sketch is completed by drawing the back lines parallel to the front lines as shown in step 3. If more of the object must be shown, the 30°–45° angle line can be increased.

The isometric pictorial of the same object is drawn in a similar way to the oblique, except that the horizontal lines on the oblique drawing become 30° lines

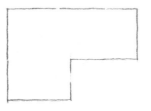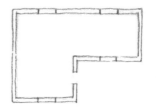

Figure 22-12

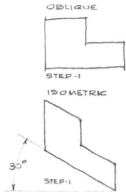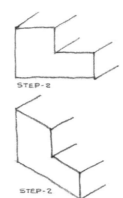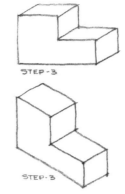

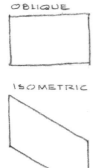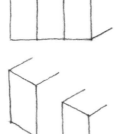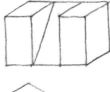

PRACTICE SKETCHING — PLANE SURFACES

Figure 22-13

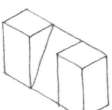

PRACTICE SKETCHING — INCLINED PLANE SURFACES

Figure 22-14

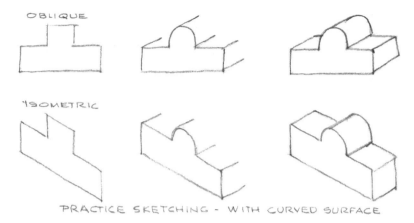

Figure 22-15

on the isometric drawing. Note that the vertical lines remain vertical in both oblique and isometric drawings. Note that the sketches are in proportion.

For practice, sketch freehand the objects' steps in Figs. 22-13 to 22-15.

Detail Sketches

The detail sketches in Fig. 22-16 are those of the box and soil sill, and a slab foundation of a basementless house. Figure 22-17 shows steps in drawing a wide-flange steel beam, a channel, and two angle irons riveted together. Figure 22-18 illustrates steps in sketching a warm air duct system.

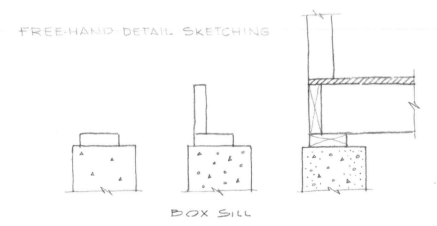

Figure 22-16

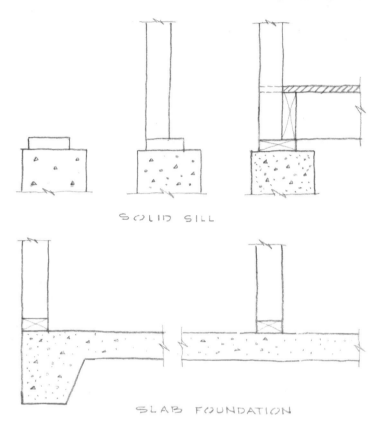

SOLID SILL

SLAB FOUNDATION

Figure 22-16 (cont.)

FREE-HAND SKETCHING STEEL MEMBERS

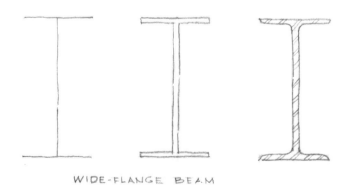

WIDE-FLANGE BEAM

Figure 22-17

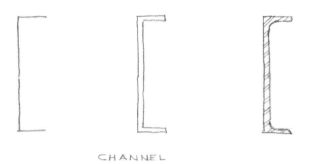

CHANNEL

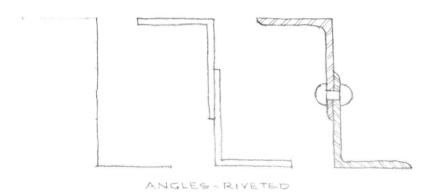

ANGLES - RIVETED

Figure 22-17 (cont.)

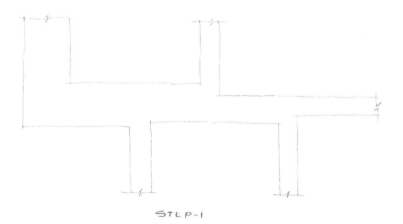

STEP-1

Figure 22-18

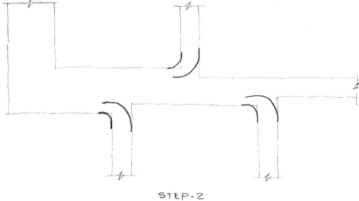

STEP-2

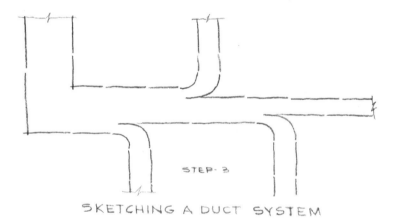

STEP-3

SKETCHING A DUCT SYSTEM

Figure 22-18 (cont.)

GLOSSARY

Anchor bolt: Used to tie down or fix a portion of a structure to a solid base below.

Area wall: An enclosed depression allowing light to enter a basement or cellar.

Asphalt: A cementitious material, dark in color, consisting of bitumens, either natural or the residue left after the distillation of crude oil.

Backfill: The earth used to fill in around exterior foundation walls.

Bearing partition: A partition that supports a vertical load.

Bearing plate: The wood member on which rest the ends of beams. A piece of steel, iron, or other material that receives the load and transmits to the masonry.

Benchmark: A fixed point of reference to determine heights and depths in surveying.

Board foot: One cubic foot of wood, such as 12 in. × 12 in. × 1 in. = 144 cu in. of wood, written as B.F. or B.M.

Branch line: A part of the plumbing piping system that drains a fixture.

Bridging: A cross bracing between joists and studs to add stiffness to floors and walls.

Btu: The abbreviation for British thermal unit—an amount of heat capable of raising one pound of water one degree Fahrenheit at sea level.

Built-up roofing: Consists of alternate layers of either asphalt- or tar-saturated felts, and hot or cold asphalt.

Casement window: A window that is hinged on the side and usually swings outward.

Cement: A mixture of lime, silica, iron oxide, and alumina, obtained by mixing an impure clay-bearing limestone with pure limestone.

Channel: A structural steel shape.

Circuit:　A path over which electric current may pass.

Circuit breaker:　An appliance for opening and closing an electric circuit.

Concrete:　A mixture of cement and aggregate (sand, gravel) made plastic with water and poured into forms.

Contract documents:　The working drawings, specifications, and other documents used for bidding and construction of a project.

Corner bead:　A metal corner strip to prevent the chipping of the plaster.

Dampproofing:　A water-resistant material applied to walls or floors to prevent moisture from seeping into the building.

Datum point:　A reference point above sea level.

Double-hung window:　A window with two sashes, one made to raise and the other to lower.

Dowel:　A metal bar embedded in the concrete footing and allowed to protrude into the concrete column or wall, to act as a tie between footing and column or wall.

Duplex outlet:　Wired for two separate circuits or may be one circuit that can supply two electrical appliances from each of the two openings.

DWV:　"Drain–waste–vent" used in plumbing.

Excavation:　The removal of all materials of every kind for the proper installation of the new foundation.

Fixture:　A receptacle attached to a plumbing system, such as a lavatory, kitchen sink, or water closet.

Flashing:　The sheet metal work over windows and doors and around chimneys to prevent leakage.

Footing:　Usually made of concrete, it is placed under the foundation wall used to distribute the imposed loads.

Formwork:　The wood, metal, or plastic shapes designed to retain the wet concrete.

Foundation:　The base or lowest part of a structure, also the footing.

Fuse:　A strip of soft metal inserted in an electric circuit, designed to melt and open the circuit should the current exceed a predetermined value.

General excavation:　The excavation for cellars or basements and floor levels below grade.

Girder:　A steel or wooden member supporting joists or beams.

Grout:　The thin mortar used for ceramic tile.

Gravel:　Detached rock particles $\frac{1}{4}$ to 3 in. in size, rounded and water-worn.

Gutter:　A trough for carrying off water.

Gypsum:　A hydrated sulfate of calcium occurring naturally in sedimentary rocks, which is used in the manufacture of plaster of paris.

Hardpan:　A coherent mixture of clay with sand, gravel, and boulders, or a cemented combination of clay, sand, and gravel.

House drain:　That part of the horizontal sewer piping inside the building receiving waste and soil from the stacks.

House sewer:　The lowest piping of the drainage system, which receives the discharge of soil, waste, and other drainage pipes of a building.

Insulation:　A special preparation (material) placed between floors, walls, and ceilings to reduce the conductivity of sound and heat.

Jamb: The side members of a window or door frame.

Joist: A small timber that supports the floor, its ends resting on walls.

Lally column: A vertical support for beams, made of iron pipe filled with concrete.

Lath: Gypsum board, expanded metal used as a base for plaster work.

Lavatory: A device used for washing face and hands.

Lime: The product resulting from the crushing of dolomite limestone, a rock consisting mainly of magnesium and calcium carbonate.

Lineal foot: A measurement of 1 ft along a straight line.

Lintel: Usually, a steel angle placed over an opening, used to support brick or other masonry over the opening.

Loose rock: Detached rock larger than gravel, generally rounded and worn as a result of having been transported by water a considerable distance from the original ledge.

Masonry: A material such as brick, stone, or concrete block used by a mason in constructing a building.

Mortar: A cement–sand–lime combination used to bond brick together.

Outlet: The point where a lamp, fixture, heater, motor, or other current-consuming device is attached to a wiring system.

Plastics: Can also be considered a cementitious material—such as adhesives, paints, and resilient flooring.

Plywood panel: Made up of an odd number of thin wood plys and a core glued together to form a panel.   Common panel size is 4 ft × 8 ft.

Public utilities: Water, gas, electricity, and sewage treatment supplied or provided by a town or city.

PVC: Polyvinyl chloride, a plastic material used in the manufacture of pipe.

Rafters: The sloped structural wooden members that support the roof.

Receptacle: A wall plug in an electrical installation.

Rock: Undisturbed, naturally formed rock masses which are part of the original rock formation.

Roll roofing: Asphalt roll roofing is manufactured in rolls, 36 in. wide and in lengths of 36 to 48 ft.   Used for roof pitches from 1 to 12 to 4 to 12.

R-value: A measurement used to calculate the resistance to the flow of heat.

Sand: Noncoherent rock particles smaller than $\frac{1}{4}$ in.

Saturated felt: An absorbent paper that has been saturated with coal tar or asphalt.   Used as a moisture barrier under floors, under wood siding, and for built-up roofs.

Septic tank: A concrete tank, placed into the ground, into which sewage is allowed to drain.

Shale: A laminated, fine textured, soft rock, composed of consolidated clay or silt.

Sheathing: The rough boarding on the outside walls or roof of a house.

Sheet piling: Where the soil is not firm, trenches may require sheathing and bracing to prevent sliding or cave-in of the soil.

Siding: The finished boarding on the outside walls of the house.

Sill cock: An outside faucet to which a hose can be attached.

Slag: A nonmetallic waste product obtained from the smelting of ferrous metals.

Slate: A dense, very fine textured, soft rock which is readily split along cleavage planes into thin sheets.

Soil stack: The vertical pipe that receives the discharge from a water closet.

Spandrel: A steel beam near the outer walls of a structure.

Square: In estimating roof surfaces, a square is 100 sq ft.

Substrate: A term to describe the material on which roofing materials are to be applied.

Terrazzo: A combination of marble chips and cement, used in floor construction, ground and polished to a high finish.

Thermostat: A switch that is activated by temperature change, used in the control of heating and air-conditioning.

Topsoil removal: If the top layer of the soil is good topsoil, it is removed by scraping the surface to a depth of about 6 in. and depositing in some corner on the site for later spreading over the area as a final topping for seeding, forming the finished grade.

Trap: A curved section of plumbing pipe under a fixture, used to trap gases from entering a room.

Trench excavation: Trenching for footing excavations is usually done with a dozer at depths indicated on the drawings.

U-factor: The number of Btu transferred in 1 hr by 1 sq ft of a building, for each 1°F temperature difference between inside and outside surfaces.   The total of all R-values of an assembly of materials.

Vapor barrier: A material such as polyethelene used to prevent moisture from passing from one material to another.

Valve: A device designed to regulate the direction of flow of fluids or gases (e.g., steam or water valve on a radiator).

Vent line: That part of the plumbing drainage system consisting of piping installed to permit adequate circulation of air in all parts of the sanitary drainage system.

Vent stack: The upper portion of a soil or waste stack above the highest fixture.

Voltage: The pressure at which an electrical system operates.   Volt is equal to the ampere times ohm, or the number of volts is found by dividing the watts by the amperes.

Water softener: A tank in which potable water is filtered through chemicals to remove minerals and thereby soften it.

Waste stack: A plumbing pipe used to receive liquid discharge from lavatories, sinks, and tubs.

INDEX

INDEX OF MATERIALS AND INSTALLATION COSTS